Integral Formulas in Riemannian Geometry

PURE AND APPLIED MATHEMATICS

A Series of Monographs

1. KENTARO YANO. Integral Formulas in Riemannian Geometry (1970)

In Preparation:

S. KOBAYASHI. Hyperbolic Spaces and Holomorphic Mappings
L. NARICI, E. BECKENSTEIN, and G. BACHMAN. Functional Analysis and Valuation Theory
V. S. VLADIMIROV. Equations of Mathematical Physics

Integral Formulas in Riemannian Geometry

KENTARO YANO

Department of Mathematics
Tokyo Institute of Technology
Tokyo, Japan

MARCEL DEKKER, INC., New York 1970

MARCEL DEKKER, INC.
95 Madison Avenue, New York, New York 10016

LIBRARY OF CONGRESS CATALOG CARD NUMBER: 74-121180

PRINTED IN THE UNITED STATES OF AMERICA

To Professor Salomon Bochner,
through whose inspiration
the main ideas
in this book developed

Preface

In classical differential geometry, integral formulas have been in vogue since Gauss, and through their use many beautiful global results have been obtained.

In Riemannian geometry, the use of integral formulas was initiated about thirty years ago by Bochner, Chern, Hodge, Thomas, and others. Integral formulas were found to be one of the very powerful tools for obtaining global results in Riemannian geometry.

The main purpose of the present book is to state some of the important results in Riemannian geometry obtained by applications of integral formulas, as well as to encourage its readers to explore this interesting branch of mathematics.

In Chapter 1 we first recall fundamental concepts and formulas in Riemannian geometry which will be used throughout the book.

In the next chapter we discuss the theory of harmonic 1-forms and Killing vector fields, which was developed mainly by Bochner, Lichnerowicz, and the present author.

Chapter 3 is devoted to the discussion of Riemannian manifolds with constant scalar curvature admitting an infinitesimal conformal transformation. The results of this chapter have been obtained by Barbance, Bishop, Couty, Goldberg, Hsiung, Kobayashi, Kurita, Lichnerowicz, Liu, Nagano, Obata, Satô, Sawaki, Tashiro, Weber, the present author, and others.

In Chapter 4 we generalize the results obtained in Chapter 2 to harmonic forms and Killing tensor fields. These results were obtained mainly by Bochner, Lichnerowicz, and the present author.

In Chapter 5 we consider hypersurfaces of a Euclidean space and prove generalizations of some of the classical theorems on closed convex surfaces. Since the remainder of the book is devoted to a discussion of hypersurfaces of a Riemannian manifold and harmonic forms and Killing vector or tensor fields in a Riemannian manifold with boundary, we recall some of the fundamental formulas in the theory of hypersurfaces of Riemannian manifolds.

After these preliminaries, we discuss, in Chapter 6, closed hypersurfaces with constant mean curvature in a Riemannian manifold. The results presented there were obtained mainly by Alexandrov, Hsiung, Katsurada, Kôjyô, Koyanagi, Liebmann, Nagai, Shahin, Süss, Tani, Tazawa, and the present author.

In Chapter 7 we study harmonic 1-forms and Killing vector fields in Riemannian manifolds with boundary, and in Chapter 8 we study harmonic forms and Killing tensor fields in such manifolds. The results of these two chapters were obtained mainly by Ako, Ba, Bochner, Duff, Hilt, Hsiung, Nakae, Spencer, Takahashi, and the present author.

Applications of integral formulas are also found in Hermitian and Kaehlerian geometries, as well as in the geometry of almost-contact manifolds. Some applications have already been stated in the author's *Differential Geometry on Complex and Almost Complex Spaces* (Pergamon Press, 1965), so these topics will not be treated in the present book.

The bibliography at the end of the book contains not only the papers and books cited in the text but also others which the author believes will be of interest to the reader. In addition, there are important papers on Hermitian and Kaehlerian geometries and the geometry of almost-contact manifolds. These are indicated by an asterisk.

Throughout the book the author will use index-free and index notation indiscriminately.

The author wishes to express his deepest gratitude to the University of Illinois and especially to Professor Samuel I. Goldberg, who invited the author to be a G. A. Miller Visiting Professor there for the first semester of the 1968–69 academic year. Professor Goldberg was kind enough to read the manuscript and to give many valuable suggestions.

For the correction of proofs the author had the valuable help of his colleagues M. Konishi, S. Kotō, Y. Kubo, M. Okumura, and S. Sawaki.

The author wishes to thank Marcel Dekker, Inc., for a most agreeable collaboration.

KENTARO YANO

Contents

CHAPTER

I

Fundamental Concepts and Formulas in Riemannian Geometry

We will recall in this chapter some fundamental concepts and formulas in Riemannian geometry to be used later. (See, e.g., Auslander and MacKenzie [1], Bishop and Crittenden [1], Bishop and Goldberg [2], Cartan [1], Chern [2, 3], Craig [1], Eisenhart [1], Flanders [1], Gerretsen [1], Guggenheimer [2], Helgason [1], Hicks [1], Hodge [1], Kobayashi and Nomizu [1, 2], Lang [1], Laugwitz [1], Levi-Civita [1], Lichnerowicz [5, 10], McConnell [1], Nomizu [1], O'Neill [1], de Rham [3], Schouten [1], Sokolnikoff [1], Spain [1], Springer [1], Sternberg [1], Struik [1], Synge and Schild [1], Thomas [2], Veblen [1], Weatherburn [1], Weber and Goldberg [1], Weyl [2], Willmore [1], Yano and Bochner [1].)

§1. Riemannian Manifolds

We consider an n-dimensional connected differentiable manifold M of class C^∞ covered by a system of coordinate neighborhoods $\{U; x^h\}$, where U denotes a neighborhood and the x^h denote local coordinates in U, with the indices $h, i, j, k, \ldots$ taking on values in the range $1, 2, \ldots, n$.

If, from any system of coordinate neighborhoods covering the manifold M, we can choose a finite number of coordinate neighborhoods which cover the whole manifold, we say that the manifold is *compact*.

If we can cover the whole manifold M by a system of coordinate neighborhoods in such a way that the Jacobian determinant $|\partial x^{h'}/\partial x^h|$ of the coordinate transformation

$$x^{h'} = x^{h'}(x^1, \ldots, x^n) \tag{1.1}$$

in every nonempty intersection of two coordinate neighborhoods $\{U; x^h\}$ and $\{U'; x^{h'}\}$ is always positive, we say that the manifold M is *orientable*.

We assume that there is given in M a positive-definite symmetric tensor field g of type (0, 2) and of differentiability class C^∞, that is, a tensor field g of type (0, 2) such that

$$g(X, Y) = g(Y, X), \qquad g(X, X) \geq 0 \tag{1.2}$$

for any vector fields X and Y, and $g(X, X) = 0$ if and only if $X = 0$. We call such a manifold M a *Riemannian manifold*.

If we denote by

$$\partial_i = \partial/\partial x^i$$

the base vectors of the so-called natural frame, we then have

$$X = X^h\,\partial_h, \qquad Y = Y^h\,\partial_h$$

for any vector fields X and Y, X^h and Y^h being local components of the vector fields X and Y, respectively, with respect to the natural frame, where we have used the so-called Einstein summation convention with respect to repeated indices.

We then have

$$g(X, Y) = g(X^j\partial_j, Y^i\partial_i) = g_{ji}X^jY^i, \tag{1.3}$$

$g_{ji} = g(\partial_j, \partial_i)$ being local components of the tensor field g. We measure the arc length of a differentiable curve

$$x^h = x^h(t) \qquad (t_1 \leq t \leq t_2),$$

in M by the definite integral

$$s = \int_{t_1}^{t_2} \sqrt{g_{ji}(x(t))\frac{dx^j}{dt}\frac{dx^i}{dt}}\,dt. \tag{1.4}$$

Thus the tensor field g is sometimes called the *fundamental metric tensor* of the Riemannian manifold M.

Since we have assumed that the quadratic form $g(X, X)$ is positive-definite, we can define the *length* or *absolute value* of a vector X by

$$|X| = \sqrt{g(X, X)}, \tag{1.5}$$

and the *angle* θ between two vectors X and Y at the same point by

$$\cos\theta = \frac{g(X, Y)}{\sqrt{g(X, X)}\sqrt{g(Y, Y)}}, \tag{1.6}$$

from which we have

$$g(X, Y) = |X| \cdot |Y| \cdot \cos\theta. \tag{1.7}$$

$g(X, Y)$ is called the *inner product* of X and Y.

The quadratic form $g(X, X)$ being positive-definite, the determinant

$$\mathscr{G} = |g_{ji}| \tag{1.8}$$

formed with local components g_{ji} of g is always positive, and consequently we can find g^{ih} such that

$$g_{ji}g^{ih} = \delta_j^h, \tag{1.9}$$

where δ_j^h is the *Kronecker delta*, that is,

$$\delta_j^h = \begin{cases} 1, & h = j, \\ 0, & h \neq j. \end{cases}$$

The g^{ih} are the *contravariant components* of the metric tensor g, g_{ji} being the *covariant components*. The δ_j^h are local components of the *identity tensor I*.

Let S be a tensor field of type (1, 2). Then $S(X, Y)$ is a vector field for any vector fields X and Y. If we put $X = X^h \partial_h$, $Y = Y^h \partial_h$, then the local components of $S(X, Y)$ with respect to the local frame are given by

$$S_{ji}{}^h X^j Y^i.$$

With this tensor field S of type (1, 2) we can associate a tensor field of type (0, 3) defined by

$$g(S(X, Y), Z),$$

X, Y, and Z being arbitrary vector fields. The value of $g(S(X, Y), Z)$ is given by

$$g_{th}S_{ji}{}^t X^j Y^i Z^h,$$

and consequently the local components of this tensor of type (0, 3) are given by

$$S_{jih} = S_{ji}{}^t g_{th}.$$

By the phrase the *lowering of index* we mean the operation which yields the local components S_{jih} of the tensor field $g(S(X, Y), Z)$ from the local components $S_{ji}{}^h$ of the tensor field $S(X, Y)$ by the formula above.

This formula is also written as

$$S_j{}^{ih} = S_{jit}g^{th},$$

which we call the *raising of index*.

When there is given a tensor field S of type (1, 2), for example, we can associate with it tensors of various types (p, q) $(p + q = 3)$ whose components are

$$S_{ji}{}^h, S_{jih}, S_j{}^{ih}, S^{jih}, \ldots$$

where

$$S_j{}^{ih} = S_{jt}{}^h g^{ti}, \qquad S^{jih} = S_t{}^{ih} g^{tj}, \ldots .$$

We say that these tensors are *conjugate* to each other, the tensors themselves being called conjugate tensors. We put

$$|S|^2 = S_{jih}S^{jih}$$

and call $|S|$ the *length* of S.

Generalizing this, we also use the notation

$$\langle S, T\rangle = S_{jih}T^{jih} \tag{1.10}$$

for two tensor fields S and T with local components S_{jih} and T^{jih}, respectively. Thus

$$\langle X, Y\rangle = g(X, Y) \tag{1.11}$$

for two vector fields and

$$\langle S, S\rangle = |S|^2 \tag{1.12}$$

for a tensor field S.

Suppose that we have a vector field X and a tensor field T of type (0, 3), for example, whose components are X^h and T_{jih}, respectively. We denote by

$$i(X)T \tag{1.13}$$

a tensor field of type (0, 2) defined by

$$X^jT_{jih}.$$

We also use the notation

$$X \lrcorner\, T \tag{1.14}$$

to denote this tensor field and

$$T \llcorner X \tag{1.15}$$

to denote the tensor field whose components are

$$T_{jih}X^h.$$

Suppose that we have a tensor field S of type $(s, 0)$ and a tensor field

T of type $(0, t)$ $(s < t)$ whose components are $S^{i_1 i_2 \cdots i_s}$ and $T_{j_1 j_2 \cdots j_t}$, respectively. We denote by

$$S \lrcorner T \tag{1.16}$$

the tensor field of type $(0, t - s)$ whose components are

$$\frac{1}{s!} S^{i_1 i_2 \cdots i_s} T_{i_1 i_2 \cdots i_s j_{s+1} \cdots j_t}$$

and by

$$T \llcorner S \tag{1.17}$$

the tensor field of type $(0, t - s)$ whose components are

$$\frac{1}{s!} T_{j_1 \cdots j_{t-s} i_1 i_2 \cdots i_s} S^{i_1 i_2 \cdots i_s}.$$

We also use the notation

$$S \otimes T \tag{1.18}$$

to denote the tensor product of S and T, that is, a tensor field of type (s, t) whose components are

$$S^{i_1 i_2 \cdots i_s} T_{j_1 j_2 \cdots j_t}.$$

We now introduce two symbols $\varepsilon_{i_1 i_2 \cdots i_n}$ and $\varepsilon^{i_1 i_2 \cdots i_n}$ defined by

$$\varepsilon_{i_1 i_2 \cdots i_n} = \varepsilon^{i_1 i_2 \cdots i_n} = \begin{cases} +1 & \text{when } (i_1, i_2, \ldots, i_n) \text{ is an} \\ & \text{even permutation of } (1, 2, \ldots, n), \\ -1 & \text{when } (i_1, i_2, \ldots, i_n) \text{ is an} \\ & \text{odd permutation of } (1, 2, \ldots, n), \\ 0 & \text{otherwise.} \end{cases} \tag{1.19}$$

The $\varepsilon_{i_1 i_2 \cdots i_n}$ are local components of a relative covariant tensor of weight -1 and $\varepsilon^{i_1 i_2 \cdots i_n}$ are those of a relative contravariant tensor of weight $+1$. These symbols are sometimes called *Eddington's epsilons*.

Since $\mathscr{G} = |g_{ji}| > 0$ is the component of a relative scalar of weight $+2$, we see that

$$e_{i_1 i_2 \cdots i_n} = \sqrt{\mathscr{G}}\, \varepsilon_{i_1 i_2 \cdots i_n}, \qquad e^{i_1 i_2 \cdots i_n} = \frac{1}{\sqrt{\mathscr{G}}} \varepsilon^{i_1 i_2 \cdots i_n} \tag{1.20}$$

are components of absolute tensors of type $(0, n)$ and $(n, 0)$, respectively. The $e_{i_1 i_2 \cdots i_n}$ and $e^{i_1 i_2 \cdots i_n}$ are both skew-symmetric and they are related by

$$e_{i_1 i_2 \cdots i_n} = e^{t_1 t_2 \cdots t_n} g_{t_1 i_1} g_{t_2 i_2} \cdots g_{t_n i_n}, \tag{1.21}$$

and

$$e^{i_1 i_2 \cdots i_n} = e_{t_1 t_2 \cdots t_n} g^{t_1 i_1} g^{t_2 i_2} \cdots g^{t_n i_n}. \tag{1.22}$$

When the Riemannian manifold M is orientable, we can define the volume element of M by

$$dV = \sqrt{\mathscr{G}}\, dx^1 \wedge dx^2 \wedge \cdots \wedge dx^n, \tag{1.23}$$

which is always positive, where the "$\wedge$" means exterior multiplication, and we can define the integral

$$\int_D f(x)\, dV \tag{1.24}$$

of a function f over the domain D of M.

Now consider a p-form

$$\omega = \frac{1}{p!}\, \omega_{i_1 i_2 \cdots i_p}\, dx^{i_1} \wedge dx^{i_2} \wedge \cdots \wedge dx^{i_p}.$$

We put

$$*\omega = \omega \mathbin{\lrcorner} e \tag{1.25}$$

and call $*\omega$ the *dual* of ω. We can easily verify that

$$**\omega = (-1)^{p(n+1)}\omega \tag{1.26}$$

and

$$\langle *\omega, *\theta \rangle = \langle \omega, \theta \rangle \tag{1.27}$$

for arbitrary p-forms ω and θ.

§2. Covariant Differentiation

Euler's differential equations of the integral

$$\int_{t_1}^{t_2} \sqrt{g_{ji}(x(t))\, \frac{dx^j}{dt} \frac{dx^i}{dt}}\, dt$$

are equations of geodesics

$$\frac{d^2 x^h}{ds^2} + \left\{ {h \atop ji} \right\} \frac{dx^j}{ds} \frac{dx^i}{ds} = 0, \tag{2.1}$$

the parameter s being an arc length of the curve, where

$$\left\{ {h \atop ji} \right\} = \tfrac{1}{2} g^{ht} (\partial_j g_{it} + \partial_i g_{jt} - \partial_t g_{ji}) \tag{2.2}$$

are *Christoffel three-index symbols.*

We denote by ∇_X the operator of covariant differentiation along the vector field X with respect to the Christoffel symbols.

So for a scalar f we have the covariant derivative $\nabla_X f$ of f along the vector field X. $\nabla_X f$ has local components

$$\nabla_X f: \quad Xf = X^i \partial_i f, \tag{2.3}$$

f and X^i being local components of f and X, respectively. $\nabla_X f$ defines a 1-form which we denote by ∇f, so that $(\nabla f)(X) = \nabla_X f$. ∇f has local components

$$\nabla f: \quad \nabla_i f = \partial_i f. \tag{2.4}$$

∇f is sometimes denoted by df.

For a vector field Y we have the covariant derivative $\nabla_X Y$ of Y along a vector field X. $\nabla_X Y$ has local components

$$\nabla_X Y: \quad X^j \nabla_j Y^h = X^j\left(\partial_j Y^h + \left\{{h \atop ji}\right\} Y^i\right), \tag{2.5}$$

X^h and Y^h being local components of X and Y, respectively. $\nabla_X Y$ defines a tensor field of type (1, 1) which we denote by ∇Y, so that $(\nabla Y)X = \nabla_X Y$. ∇Y has local components

$$\nabla Y; \quad \nabla_j Y^h = \partial_j Y^h + \left\{{h \atop ji}\right\} Y^i. \tag{2.6}$$

We note here that

$$\nabla_X Y - \nabla_Y X - [X, Y] = 0,$$

which means that the connection ∇ has no torsion, where $[X, Y]$ represents a vector field defined by

$$[X, Y]f = X(Yf) - Y(Xf) \tag{2.7}$$

for an arbitrary function f, and has local components

$$[X, Y]: \quad X^i \partial_i Y^h - Y^i \partial_i X^h, \tag{2.8}$$

X^h and Y^h being local components of X and Y, respectively.

For a 1-form ω, we have the covariant derivative $\nabla_X \omega$ along X, which is a 1-form defined by

$$(\nabla_X \omega)(Y) = \nabla_X(\omega(Y)) - \omega(\nabla_X Y) \tag{2.9}$$

for any vector fields X and Y. $\nabla_X \omega$ has local components

$$\nabla_X \omega: \quad X^j \nabla_j \omega_i = X^j\left(\partial_j \omega_i - \left\{{h \atop ji}\right\}\omega_h\right), \tag{2.10}$$

ω_i and X^h being local components of ω and X, respectively. $\nabla_X\omega$ defines a tensor field of type (0, 2) which we denote by $\nabla\omega$, so that $(\nabla\omega)(X, Y) = (\nabla_X\omega)(Y)$. $\nabla\omega$ has local components

$$\nabla\omega: \quad \nabla_j\omega_i = \partial_j\omega_i - \begin{Bmatrix} h \\ ji \end{Bmatrix}\omega_h. \tag{2.11}$$

We note here that

$$(\nabla\omega)(X, Y) - (\nabla\omega)(Y, X) \tag{2.12}$$

has the local components

$$\partial_j\omega_i - \partial_i\omega_j$$

and is therefore independent of Christoffel symbols.

For a general tensor field, say, S of type (1, 2), we have the covariant derivative $\nabla_X S$ along a vector field X, which is a tensor field of type (1, 2) and is defined by

$$(\nabla_X S)(Y, Z) = \nabla_X(S(Y, Z)) - S(\nabla_X Y, Z) - S(Y, \nabla_X Z) \tag{2.13}$$

for any vector fields X, Y, and Z. $\nabla_X S$ has the local components

$$\nabla_X S: \quad X^k\nabla_k S_{ji}{}^h = X^k\left(\partial_k S_{ji}{}^h + \begin{Bmatrix} h \\ kt \end{Bmatrix} S_{ji}{}^t - \begin{Bmatrix} t \\ kj \end{Bmatrix} S_{ti}{}^h - \begin{Bmatrix} t \\ ki \end{Bmatrix} S_{jt}{}^h\right), \tag{2.14}$$

$S_{ji}{}^h$ and X^h being local components of S and X, respectively. $\nabla_X S$ defines a tensor field of type (1, 3) which we denote by ∇S, so that $(\nabla S)(X, Y, Z) = (\nabla_X S)(Y, Z)$. ∇S has the local components

$$\nabla S: \quad \nabla_k S_{ji}{}^h = \partial_k S_{ji}{}^h + \begin{Bmatrix} h \\ kt \end{Bmatrix} S_{ji}{}^t - \begin{Bmatrix} t \\ kj \end{Bmatrix} S_{ti}{}^h - \begin{Bmatrix} t \\ ki \end{Bmatrix} S_{jt}{}^h. \tag{2.15}$$

We can also define the covariant derivative of relative tensor fields. For example, the covariant derivative of a relative scalar $\mathscr{G}$ of weight $+2$ is given by

$$\nabla_i\mathscr{G} = \partial_i\mathscr{G} - 2\begin{Bmatrix} t \\ it \end{Bmatrix}\mathscr{G}.$$

We now have

$$\begin{aligned} \nabla g = 0: \quad & \nabla_k g_{ji} = \partial_k g_{ji} - \begin{Bmatrix} t \\ kj \end{Bmatrix} g_{ti} - \begin{Bmatrix} t \\ ki \end{Bmatrix} g_{jt} = 0, \\ & \nabla_k g^{ih} = \partial_k g^{ih} + \begin{Bmatrix} i \\ kt \end{Bmatrix} g^{th} + \begin{Bmatrix} h \\ kt \end{Bmatrix} g^{it} = 0, \end{aligned} \tag{2.16}$$

$$\nabla\mathscr{G} = 0: \quad \nabla_i\mathscr{G} = \partial_i\mathscr{G} - 2\begin{Bmatrix} t \\ it \end{Bmatrix}\mathscr{G} = 0, \tag{2.17}$$

$$\nabla I = 0: \quad \nabla_j \delta_i^h = \partial_j \delta_i^h + \begin{Bmatrix} h \\ jt \end{Bmatrix} \delta_i^t - \begin{Bmatrix} t \\ ji \end{Bmatrix} \delta_t^h = 0, \tag{2.18}$$

$$\begin{aligned} \nabla e = 0: \quad \nabla_j e_{i_1 i_1 \cdots i_n} &= \partial_j e_{i_1 i_2 \cdots i_n} - \begin{Bmatrix} t \\ j i_1 \end{Bmatrix} e_{t i_2 \cdots i_n} \\ &\qquad - \cdots - \begin{Bmatrix} t \\ j i_n \end{Bmatrix} e_{i_1 i_2 \cdots t} = 0, \\ \nabla_j e^{i_1 i_2 \cdots i_n} &= \partial_j e^{i_1 i_2 \cdots i_n} + \begin{Bmatrix} i_1 \\ jt \end{Bmatrix} e^{t i_2 \cdots i_n} \\ &\qquad + \cdots + \begin{Bmatrix} i_n \\ jt \end{Bmatrix} e^{i_1 i_2 \cdots t} = 0. \end{aligned} \tag{2.19}$$

It is known that the connection defined by the Christoffel symbols is the unique connection which has no torsion and satisfies $\nabla g = 0$. We call this connection the *Riemannian connection*.

We now consider a p-form

$$\omega = \frac{1}{p!} \omega_{i_1 i_2 \cdots i_p} dx^{i_1} \wedge dx^{i_2} \wedge \cdots \wedge dx^{i_p}$$

or a skew-symmetric tensor field of type $(0, p)$. The $\omega_{i_1 i_2 \cdots i_p}$ are local components of the p-form or the skew-symmetric tensor field of type $(0, p)$.

The *exterior differential* or simply *differential* $d\omega$ of ω is the $(p + 1)$-form with the local expression

$$\begin{aligned} d\omega = \frac{1}{(p+1)!} (\partial_i \omega_{i_1 i_2 \cdots i_p} - \partial_{i_1} \omega_{i i_2 \cdots i_p} - \partial_{i_2} \omega_{i_1 i i_3 \cdots i_p} \\ - \cdots - \partial_{i_p} \omega_{i_1 i_2 \cdots i}) \, dx^i \wedge dx^{i_1} \wedge \cdots \wedge dx^{i_p}. \end{aligned} \tag{2.20}$$

It can also be written as

$$\begin{aligned} d\omega = \frac{1}{(p+1)!} (\nabla_i \omega_{i_1 i_2 \cdots i_p} - \nabla_{i_1} \omega_{i i_2 \cdots i_p} - \nabla_{i_2} \omega_{i_1 i i_3 \cdots i_p} \\ - \cdots - \nabla_{i_p} \omega_{i_1 i_2 \cdots i}) \, dx^i \wedge dx^{i_1} \wedge \cdots \wedge dx^{i_p}. \end{aligned} \tag{2.21}$$

It is well known that

$$d(d\omega) = 0 \tag{2.22}$$

for any p-form ω.

From a p-form

$$\omega = \frac{1}{p!}\,\omega_{i_1 i_2 \cdots i_p}\, dx^{i_1} \wedge dx^{i_2} \wedge \cdots \wedge dx^{i_p},$$

we construct a $(p-1)$-form

$$\delta\omega = -\frac{1}{(p-1)!}\,(g^{ji}\nabla_j \omega_{i i_2 \cdots i_p})\, dx^{i_2} \wedge \cdots \wedge dx^{i_p} \tag{2.23}$$

and call it the *codifferential* of the p-form ω. If ω is a function, we put $\delta f = 0$.

It can be verified that

$$\delta(\delta\omega) = 0 \tag{2.24}$$

for any p-form ω.

We also define the codifferential δS of the more general tensor field S. For example, let S be a tensor field of type (0, 3) and denote by S_{jih} its local components. Then δS is the tensor field of type (0, 2):

$$\delta S: \quad -g^{kj}\nabla_k S_{jih} \tag{2.25}$$

or

$$\delta S: \quad -\nabla^j S_{jih}, \tag{2.26}$$

where

$$\nabla^j = g^{jt}\nabla_t.$$

A p-form ω, or a skew-symmetric tensor field of type $(0, p)$, is said to be *harmonic*[1] if it satisfies

$$d\omega = 0 \quad \text{and} \quad \delta\omega = 0. \tag{2.27}$$

If we put

$$\Delta = \delta d + d\delta, \tag{2.28}$$

then we have

$$\Delta\omega = 0 \tag{2.29}$$

for a harmonic p-form ω.

If we have a vector field X with local components X^h, then there is associated with it a 1-form

$$\xi = g_{ji}X^j\, dx^i = X_i\, dx^i.$$

[1] This is the definition given by W. V. D. Hodge as opposed to $\Delta\omega = 0$ as defined by K. Kodaira. However, for compact and orientable manifolds, the two definitions are equivalent.

The codifferential $\delta\xi$ of ξ is given by

$$\delta\xi = -\nabla_i X^i = -g^{ji}\nabla_j X_i. \tag{2.30}$$

This will also be denoted by δX.

We now state

GREEN'S THEOREM. *In a compact orientable Riemannian manifold M without boundary, we have*

$$\int_M (\delta X)\, dV = 0$$

for any vector field X, or

$$\int_M (\nabla_i X^i)\, dV = 0 \quad or \quad \int_M (g^{ji}\nabla_j X_i)\, dV = 0,$$

X^i, X_i being local components of X.

If we have a function f, we can form the codifferential δdf of the differential df of f. This has the local expression $-g^{ji}\nabla_j\nabla_i f$. We put

$$\Delta f = -g^{ji}\nabla_j\nabla_i f \tag{2.31}$$

and call the operator Δ the *Laplacian*.[1]

Applying Green's theorem, we have

THEOREM 2.1. *In a compact and orientable Riemannian manifold M without boundary, we have*

$$\int_M \Delta f\, dV = 0, \tag{2.32}$$

for any function f of M.

The existence of harmonic forms in M is closely related to the topology of M. In fact we have

THEOREM OF HODGE [1]. *In a compact and orientable Riemannian manifold, the number of linearly independent (with constant real coefficients) harmonic p-forms is equal to the p-th-dimensional Betti number B_p of the manifold.*

§3. Curvature Tensor

Let X, Y, and Z be three arbitrary vector fields in M. Computing

$$\nabla_X\nabla_Y Z - \nabla_Y\nabla_X Z - \nabla_{[X,Y]}Z,$$

[1] The Laplacian so defined differs in sign from the usual Laplacian of a function.

we find that this expression defines a tensor field of type (1, 3) and consequently we put

$$\nabla_X\nabla_Y Z - \nabla_Y\nabla_X Z - \nabla_{[X,Y]}Z = K(X, Y)Z, \tag{3.1}$$

$K(X, Y)$ being a tensor field of type (1, 1) which is linear in X and Y.

With local components the equation above can be written as

$$\nabla_k\nabla_j Z^h - \nabla_j\nabla_k Z^h = K_{kji}{}^h Z^i, \tag{3.2}$$

where

$$K_{kji}{}^h = \partial_k\left\{ {h \atop ji} \right\} - \partial_j\left\{ {h \atop ki} \right\} + \left\{ {h \atop kt} \right\}\left\{ {t \atop ji} \right\} - \left\{ {h \atop jt} \right\}\left\{ {t \atop ki} \right\}$$

are the local components of the tensor field of type (1, 3). This tensor is called the *Riemann-Christoffel curvature tensor* of the Riemannian manifold M.

A necessary and sufficient condition for M to be covered by a system of coordinate neighborhoods $\{U; x^h\}$ such that the components of the fundamental metric tensor are all constant in any of the coordinate neighborhoods is that the Riemann-Christoffel curvature tensor vanishes identically. Such a Riemannian manifold is said to be *locally Euclidean*, or *locally flat*.

We also have

$$(\nabla_X\nabla_Y\omega)(Z) - (\nabla_Y\nabla_X\omega)(Z) - (\nabla_{[X,Y]}\omega)(Z) = -\omega(K(X, Y)Z) \tag{3.3}$$

or

$$\nabla_k\nabla_j\omega_i - \nabla_j\nabla_k\omega_i = -K_{kji}{}^h\omega_h \tag{3.4}$$

for a 1-form ω with local components ω_i and

$$\begin{aligned}(\nabla_X\nabla_Y S)(Z, U) &- (\nabla_Y\nabla_X S)(Z, U) - (\nabla_{[X,Y]}S)(Z, U)\\ &= K(X, Y)S(Z, U) - S(K(X, Y)Z, U) - S(Z, K(X, Y)U)\end{aligned} \tag{3.5}$$

or

$$\nabla_l\nabla_k S_{ji}{}^h - \nabla_k\nabla_l S_{ji}{}^h = K_{lkt}{}^h S_{ji}{}^t - K_{lkj}{}^t S_{ti}{}^h - K_{lki}{}^t S_{jt}{}^h \tag{3.6}$$

for a tensor field S of type (1, 2) with local components $S_{ji}{}^h$, where X, Y, Z, U are arbitrary vector fields.

The formulas above are called *Ricci identities*.

It can be easily verified that the curvature tensor K satisfies

$$K(X, Y) + K(Y, X) = 0, \tag{3.7}$$

$$K(X, Y)Z + K(Y, Z)X + K(Z, X)Y = 0, \tag{3.8}$$

or

$$K_{kji}{}^{h} + K_{jki}{}^{h} = 0, \tag{3.9}$$

$$K_{kji}{}^{h} + K_{jik}{}^{h} + K_{ikj}{}^{h} = 0 \tag{3.10}$$

in terms of local coordinates.

The tensor field K of type (0, 4) defined by

$$K(X, Y; U, V) = g(K(X, Y)U, V) \tag{3.11}$$

has local components

$$K_{kjih} = K_{kji}{}^{t} g_{th},$$

which are called *covariant components* of the curvature tensor.

The curvature tensor K satisfies

$$K(X, Y; U, V) + K(Y, X; U, V) = 0 \tag{3.12}$$

$$K(X, Y; U, V) + K(Y, U; X, V) + K(U, X; Y, V) = 0, \tag{3.13}$$

$$K(X, Y; U, V) + K(X, Y; V, U) = 0, \tag{3.14}$$

$$K(X, Y; U, V) = K(U, V; X, Y) \tag{3.15}$$

or

$$K_{kjih} + K_{jkih} = 0, \tag{3.16}$$

$$K_{kjih} + K_{jikh} + K_{ikjh} = 0, \tag{3.17}$$

$$K_{kjih} + K_{kjhi} = 0, \tag{3.18}$$

$$K_{kjih} = K_{ihkj} \tag{3.19}$$

in terms of local components.

The curvature tensor also satisfies the so-called Bianchi identity.

$$(\nabla_X K)(Y, Z) + (\nabla_Y K)(Z, X) + (\nabla_Z K)(X, Y) = 0, \tag{3.20}$$

or

$$\nabla_l K_{kji}{}^{h} + \nabla_k K_{jli}{}^{h} + \nabla_j K_{lki}{}^{h} = 0, \tag{3.21}$$

$$\nabla_l K_{kjih} + \nabla_k K_{jlih} + \nabla_j K_{lkih} = 0 \tag{3.22}$$

in terms of local coordinates.

In terms of the n local orthonormal vector fields $X_1, X_2, \ldots, X_n$,

$$R(Y, Z) = \sum_{i=1}^{n} K(X_i, Y; Z, X_i) \tag{3.23}$$

defines a global tensor field R of type (0, 2) with local components

$$K_{ji} = K_{tji}{}^{t} = g^{ts} K_{tjis}. \tag{3.24}$$

Using the relations satisfied by the curvature tensor K, we can easily prove that the tensor $R(Y, Z)$ is symmetric in Y and Z, that is,

$$R(Y, Z) = R(Z, Y), \tag{3.25}$$

or

$$K_{ji} = K_{ij}. \tag{3.26}$$

The tensor field R is called the *Ricci tensor*.

Again, in terms of the n local orthonormal vector fields $X_1, X_2, \ldots, X_n$, we can define a global scalar field

$$r = \sum_{i=1}^{n} R(X_i, X_i) \tag{3.27}$$

with the local component

$$r = g^{ji}K_{ji}. \tag{3.28}$$

The scalar r is called the *scalar curvature* of the Riemannian manifold M.

From the Bianchi identity

$$\nabla_l K_{kji}{}^h + \nabla_k K_{jli}{}^h + \nabla_j K_{lki}{}^h = 0,$$

we can deduce, by contraction with respect to l and h,

$$\nabla_t K_{kji}{}^t - \nabla_k K_{ji} + \nabla_j K_{ki} = 0$$

from which, by transvection with g^{ji},

$$\nabla_t K_k{}^t = \tfrac{1}{2}\nabla_k r, \tag{3.29}$$

where

$$K_k^h = K_{kj}g^{jh} = g^{ts}K_{kts}{}^h. \tag{3.30}$$

The tensor field of type (1, 1) having the local components K_k^h is denoted by Q; that is,

$$g(QX, Y) = R(X, Y) \tag{3.31}$$

for any vector fields X and Y.

For a surface in the 3-dimensional Euclidean space regarded as a 2-dimensional Riemannian manifold, the Gaussian curvature k is given by

$$k = -\frac{K_{1212}}{g_{11}g_{22} - g_{12}g_{21}}. \tag{3.32}$$

We consider two linearly independent vectors X and Y at a point P and a 2-dimensional submanifold generated by geodesics which pass through the point P and are tangent to the plane γ determined by the two vectors X and Y at P. Then the Gaussian curvature k of this 2-dimensional submanifold at the point P is given by

$$k = -\frac{K(X, Y; X, Y)}{g(X, X)g(Y, Y) - g(X, Y)^2}. \tag{3.33}$$

It is easily seen that this k is uniquely determined by the plane γ and independent of the choice of X and Y on it. The number k is called the *sectional curvature* of M at P with respect to the *section* γ.

Suppose that the sectional curvature at P is independent of the section; then we have

$$K(X, Y; X, Y) = -k\{g(X, X)g(Y, Y) - g(X, Y)^2\} \tag{3.34}$$

for any vectors X and Y at P, or

$$K_{kjih} = k(g_{kh}g_{ji} - g_{jh}g_{ki}), \tag{3.35}$$

or

$$K_{kji}{}^h = k(\delta_k^h g_{ji} - \delta_j^h g_{ki}) \tag{3.36}$$

at P in terms of local components. The last equation can be written as

$$K(X, Y)Z = k\{g(Y, Z)X - g(X, Z)Y\} \tag{3.37}$$

at P. If this is the case at every point of the manifold M, then the curvature tensor has the above form at every point of the manifold, k being in general a function on M. But if $n > 2$, we can prove, using the Bianchi identity, that k is a constant. If $n = 2$, the curvature tensor always has the above form, with k not necessarily constant.

The equation above can also be written as

$$K_{kji}{}^h = \frac{r}{n(n-1)}(\delta_k^h g_{ji} - \delta_j^h g_{ki}), \tag{3.38}$$

with r being the scalar curvature. A Riemannian manifold is said to be of *constant (sectional) curvature* when the sectional curvature is independent of the section and is equal to a constant. The curvature tensor of a space of constant curvature has the above form where k is a constant.

Now take a vector X at a fixed point P of the manifold M and consider $n - 1$ vectors $X_2, X_3, \ldots, X_n$ at P which are orthogonal to X and also mutually orthogonal. Then the sum of sectional curvatures at P with respect to the sections determined by (X, X_2), $(X, X_3), \ldots, (X, X_n)$, respectively, is given by

$$\lambda = \frac{R(X, X)}{g(X, X)}$$

and is independent of the choice of the $n - 1$ vectors $X_2, \ldots, X_n$. This is called the *Ricci curvature* at P with respect to the vector X.

If the Ricci curvature at P is independent of the vector X, then the Ricci tensor must be of the form

$$R = \lambda g, \tag{3.39}$$

that is,

$$K_{ji} = \lambda g_{ji} \tag{3.40}$$

at P in terms of local coordinates. If this is the case at every point of the manifold, then the Ricci tensor has the above form at every point of the manifold, with λ being in general a function on M. If $n > 2$, we can prove by the Bianchi identity that λ is a constant. If $n = 2$, the Ricci tensor always has the above form, with λ not necessarily constant.

The equation above can also be written as

$$K_{ji} = \frac{r}{n} g_{ji}, \tag{3.41}$$

with r being the scalar curvature. A Riemannian manifold whose Ricci tensor is of the above form is called an *Einstein space*.

In general the scalar curvature r has the following geometrical interpretation. Consider n local orthonormal vector fields $X_1, X_2, \ldots, X_n$ at a point P of the manifold. Then the sum of Ricci curvatures

$$R(X_1, X_1) + \cdots + R(X_n, X_n)$$

is independent of the choice of $X_1, \ldots, X_n$ and is equal to the scalar curvature at P.

The following theroem is well-known:

THEOREM (Gauss-Bonnet). *Let M be an n-dimensional (n even), compact and orientable Riemannian manifold and let $\chi(M)$ be the Euler-Poincaré characteristic of M. Then*

$$\chi(M) = \frac{n!}{2^n \pi^{n/2}\left(\frac{n}{2}!\right)} \int_M e^{i_1 i_2 \cdots i_{n-1} i_n} e^{j_1 j_2 \cdots j_{n-1} j_n} \times K_{i_1 i_2 j_1 j_2} K_{i_3 i_4 j_3 j_4} \cdots K_{i_{n-1} i_n j_{n-1} j_n}\, dV. \tag{3.42}$$

§4. Projective and Conformal Transformations

We consider an n-dimensional differentiable manifold M of class C^∞ and assume that there is given in M an affine connection ∇ without torsion whose local parameters are Γ^h_{ji}. Then the so-called *paths* or *autoparallel curves* are defined as curves $x^h = x^h(t)$ which satisfy the system of differential equations

$$\frac{d^2x^h}{dt^2} + \Gamma^h_{ji}\frac{dx^j}{dt}\frac{dx^i}{dt} = \alpha(t)\frac{dx^h}{dt}, \tag{4.1}$$

$\alpha(t)$ being a function defined along curves, or

$$\frac{d^2x^h}{ds^2} + \Gamma^h_{ji}\frac{dx^j}{ds}\frac{dx^i}{ds} = 0, \tag{4.2}$$

s being the so-called *affine parameter* on paths.

If there is given another connection ∇^* without torsion whose local parameters are Γ^{*h}_{ji}, and if the affine connections ∇^* and ∇ give the same system of paths, we must have

$$H(X, Y) = \pi(X)Y + \pi(Y)X, \tag{4.3}$$

where

$$H(X,Y) = (\nabla^* - \nabla)(X,Y) = (\nabla^*_X - \nabla_X)Y, \tag{4.4}$$

π being a certain 1-form, or

$$\Gamma^{*h}_{ji} - \Gamma^h_{ji} = \delta^h_j p_i + \delta^h_i p_j, \tag{4.5}$$

where the p_i are the local components of π. The change of ∇ given by the equation above is called a *projective change* of ∇. We also say that ∇^* and ∇ are *projectively related.*

Denoting by K^* and K the curvature tensors of ∇^* and ∇, respectively, we have

$$K^*(X, Y)Z = K(X, Y)Z - p(Y, Z)X + p(X, Z)Y + \{p(X, Y) - p(Y, X)\}Z, \tag{4.6}$$

where

$$p(X, Y) = (\nabla_X\pi)(Y) - \pi(X)\pi(Y), \tag{4.7}$$

or

$$K^{*}_{kji}{}^{h} = K_{kji}{}^{h} - \delta^h_k p_{ji} + \delta^h_j p_{ki} + (p_{kj} - p_{jk})\delta^h_i, \tag{4.8}$$

where

$$p_{ji} = \nabla_j p_i - p_j p_i. \tag{4.9}$$

If we put

$$L(X, Y) = \frac{1}{n^2 - 1}(-nR(X, Y) + R(Y, X)) \tag{4.10}$$

and define $L^*(X, Y)$ similarly, we have, from the above equation,

$$L^*(X, Y) = L(X, Y) + p(X, Y) \tag{4.11}$$

or

$$P^*_{ji} = P_{ji} + p_{ji}, \tag{4.12}$$

P^*_{ji} and P_{ji} being local components of L^* and L, respectively.

Eliminating p_{ji}, we find

$$P^*(X, Y)Z = P(X, Y)Z, \tag{4.13}$$

where

$$P(X, Y)Z = K(X, Y)Z + L(Y, Z)X - L(X, Z)Y - \{L(X, Y) - L(Y, X)\}Z, \tag{4.14}$$

P^* having a similar definition, or

$$P^*_{kji}{}^h = P_{kji}{}^h, \tag{4.15}$$

where

$$P_{kji}{}^h = K_{kji}{}^h + \delta^h_k P_{ji} - \delta^h_j P_{ki} - (P_{kj} - P_{jk})\delta^h_i, \tag{4.16}$$

$P^*_{kji}{}^h$ having a similar expression. These equations show that the tensor P is invariant under a projective change of ∇. The tensor P is called the *Weyl projective curvature tensor*. The Weyl projective curvature tensor vanishes identically for $n = 2$ (Weyl [1, 3]).

We also have

$$Q^*(X, Y, Z) = Q(X, Y, Z) + \pi(P(X, Y)Z), \tag{4.17}$$

where

$$Q(X, Y, Z) = (\nabla_X L)(Y, Z) - (\nabla_Y L)(X, Z), \tag{4.18}$$

$Q^*(X, Y, Z)$ having a similar definition, or

$$P^*_{kji} = P_{kji} + P_{kji}{}^h p_h, \tag{4.19}$$

where

$$P_{kji} = \nabla_k P_{ji} - \nabla_j P_{ki} \tag{4.20}$$

in local coordinates, P^*_{kji} having a similar expression.

If the connection ∇ is projectively related to a connection ∇^* which is affinely flat, that is, whose curvature tensor vanishes, then the manifold with the affine connection ∇ is said to be *projectively flat*.

The following theorem of Weyl is well-known:

THEOREM 4.1. *A necessary and sufficient condition for a manifold with a symmetric affine connection ∇ to be projectively flat is that*

$$P = 0 \quad \textit{for} \quad n > 2$$

and

$$Q = 0 \quad for \quad n = 2.$$

We can also prove

THEOREM 4.2. *A projectively flat Riemannian manifold is of constant curvature.*

We next consider an n-dimensional Riemannian manifold M with the fundamental metric tensor g. The change of the metric

$$g^* = \rho^2 g, \tag{4.21}$$

where ρ is a certain positive function, does not change the angle between two vectors at a point and so is called a *conformal transformation* of the metric. If ρ is a positive constant, the conformal transformation is said to be *homothetic*.

Denoting by ∇^* the operator of covariant differentiation with respect to the Christoffel symbols $\left\{ {h \atop ji} \right\}^*$ formed with g^*, we have

$$(\nabla^* - \nabla)(X, Y) = \omega(X)Y + \omega(Y)X - g(X, Y)U \tag{4.22}$$

for any vector fields X, Y, where ω is a 1-form defined by

$$\omega = d \log \rho$$

and U is a vector field defined by

$$g(U, X) = \omega(X),$$

or

$$\left\{ {h \atop ji} \right\}^* - \left\{ {h \atop ji} \right\} = \delta_j^h \rho_i + \delta_i^h \rho_j - g_{ji} \rho^h, \tag{4.23}$$

where

$$\rho_i = \nabla_i \log \rho, \qquad \rho^h = \rho_i g^{ih}$$

in local coordinates.

Denoting by K^* the curvature tensor of the Riemannian metric g^*, we have

$$K^*(X, Y)Z = K(X, Y)Z - s(Y, Z)X \\ + s(X, Z)Y - g(Y, Z)SX + g(X, Z)SY \tag{4.24}$$

for any vector fields X, Y, Z, where s is the tensor field of type (0, 2) defined by

$$s(X, Y) = (\nabla_X \omega)(Y) - \omega(X)\omega(Y) + \tfrac{1}{2}\omega(U)g(X, Y) \tag{4.25}$$

and S is the tensor field of type (1,1) defined by

$$g(SX, Y) = s(X, Y),$$

or

$$K^{*}_{kji}{}^{h} = K_{kji}{}^{h} - \delta^h_k \rho_{ji} + \delta^h_j \rho_{ki} - \rho_k{}^h g_{ji} + \rho_j{}^h g_{ki}, \tag{4.26}$$

where

$$\rho_{ji} = \nabla_j \rho_i - \rho_j \rho_i + \tfrac{1}{2}\rho_t \rho^t g_{ji}, \qquad \rho_k{}^h = \rho_{kt} g^{th} \tag{4.27}$$

in local coordinates.

From the equation above, we deduce

$$L^* = L + s, \tag{4.28}$$

where

$$L(X, Y) = -\frac{1}{n-2} R(X, Y) + \frac{1}{2(n-1)(n-2)} rg(X, Y), \tag{4.29}$$

L^* having a similar definition, or

$$C^*_{ji} = C_{ji} + \rho_{ji}, \tag{4.30}$$

where

$$C_{ji} = -\frac{1}{n-2} K_{ji} + \frac{1}{2(n-1)(n-2)} rg_{ji}, \tag{4.31}$$

C^*_{ji} having a similar expression.

Thus, eliminating s, we find

$$C^* = C, \tag{4.32}$$

where

$$\begin{aligned} C(X, Y)Z = K(X, Y)Z &+ L(Y, Z)X \\ &- L(X, Z)Y + g(Y, Z)NX - g(X, Z)NY \end{aligned} \tag{4.33}$$

and

$$g(NX, Y) = L(X, Y) \tag{4.34}$$

for any vector fields X, Y, Z, and C^* has a similar expression, or

$$C^{*}_{kji}{}^{h} = C_{kji}{}^{h}, \tag{4.35}$$

where

$$C_{kji}{}^{h} = K_{kji}{}^{h} + \delta^h_k C_{ji} - \delta^h_j C_{ki} + C_k{}^h g_{ji} - C_j{}^h g_{ki}, \tag{4.36}$$

$$C_k{}^h = C_{kt} g^{th},$$

and $C^{*}_{kji}{}^{h}$ has a similar expression.

Thus the tensor field C is invariant under any conformal transformation of the metric. It is called the *Weyl conformal curvature tensor*. The tensor C vanishes identically for $n = 3$ (Weyl [1, 3]).

We also have

$$D^*(X, Y, Z) = D(X, Y, Z) + \omega(C(X, Y)Z), \tag{4.37}$$

where

$$D(X, Y, Z) = (\nabla_X L)(Y, Z) - (\nabla_Y L)(X, Z), \tag{4.38}$$

D^* having a similar expression, or

$$C^*_{kji} = C_{kji} + C_{kji}{}^h \rho_h \tag{4.39}$$

in local coordinates, where

$$C_{kji} = \nabla_k C_{ji} - \nabla_j C_{ki}, \tag{4.40}$$

C^*_{kji} having a similar expression.

If a Riemannian metric g is conformally related to a Riemannian metric g^* which is locally Euclidean, then the Riemannian manifold with the metric g is said to be *conformally flat*.

The following theorem of Weyl is well-known.

THEOREM 4.3 (Weyl [1, 3]). *A necessary and sufficient condition for a Riemannian manifold to be conformally flat is that*

$$C = 0 \quad \textit{for} \quad n > 3$$

and

$$D = 0 \quad \textit{for} \quad n = 3.$$

Yamabe proved the following

THEOREM 4.4[1] (Yamabe [1]). *For an arbitrary Riemannian metric on an n-dimensional ($n \geq 3$) compact manifold of class C^∞, there always exists a Riemannian metric which is conformal to the given Riemannian metric and whose scalar curvature is a constant.*

§5. Lie Derivatives (Yano [9])

We consider a point transformation

$$\pi: \quad P \to {}'P \tag{5.1}$$

[1] Yamabe's theorem appears to be in doubt (see N. S. Trudinger, Remarks concerning the conformal deformation of Riemannian structures on compact manifolds, *Ann. Scuola Normale Sup. Pisa* **22** (1968), 265–275). However, the following is known. For $n \geq 3$, a structure of positive constant scalar curvature may be topologically deformed into one of constant negative scalar curvature, so the sign of the scalar curvature has no global significance in more than two dimensions.

in M, or, in local coordinates,

$$\pi: \quad x^h \to {}'x^h = f^h(x). \tag{5.2}$$

Suppose that we have a geometric object field $\Omega(P)$ and we bring back the object $\Omega('P)$ at $'P$ to P by (the differential of) the transformation inverse to π; then we have a geometric object $'\Omega(P)$ at P. We call the difference

$$'\Omega(P) - \Omega(P) \tag{5.3}$$

the *Lie difference* of Ω with respect to the point transformation under consideration.

If we have a one-parameter group of transformations $'x^h = f^h(x, t)$ generated by a vector field X, we define

$$\pounds_X \Omega = \lim_{t \to 0} \frac{1}{t} ('\Omega(P) - \Omega(P)) \tag{5.4}$$

as the *Lie derivative* of the object Ω with respect to X, t being the so-called canonical parameter.

According to this definition, the Lie derivative of a function f with respect to X with local components X^h is given by

$$\pounds_X f = Xf = X^i \, \partial_i f. \tag{5.5}$$

The Lie derivative of a vector field Y with respect to X is given by

$$\pounds_X Y = [X, Y], \tag{5.6}$$

or, in a local coordinate system,

$$(\pounds_X Y)^h = X^i \, \partial_i Y^h - Y^i \, \partial_i X^h. \tag{5.7}$$

The Lie derivative of a 1-form ω with respect to X is given by

$$(\pounds_X \omega)(Y) = X(\omega(Y)) - \omega([X, Y]) \tag{5.8}$$

or, in a local coordinate system,

$$(\pounds_X \omega)_i = X^j \, \partial_j \omega_i + \omega_j \, \partial_i X^j. \tag{5.9}$$

The Lie derivative of a tensor field, say, S of type $(1, 2)$, with respect to X is given by

$$(\pounds_X S)(Y, Z) = [X, S(Y, Z)] - S([X, Y], Z) - S(Y, [X, Z]), \tag{5.10}$$

or, in a local coordinate system,

$$(\pounds_X S)_{ji}{}^h = X^k \, \partial_k S_{ji}{}^h - S_{ji}{}^t \, \partial_t X^h + S_{ti}{}^h \, \partial_j X^t + S_{jt}{}^h \, \partial_i X^t. \tag{5.11}$$

The Lie derivative $\pounds_X \nabla$ of a symmetric affine connection ∇ with respect to a vector field X is a tensor field of type (1, 2) whose local components are given by

$$(\pounds_X \nabla)_{ji}{}^h = \nabla_j \nabla_i X^h + K_{kji}{}^h X^k. \tag{5.12}$$

We sometimes denote $(\pounds_X Y)^h$, $(\pounds_X \omega)_i$, $(\pounds_X S)_{ji}{}^h$, $(\pounds_X \nabla)_{ji}{}^h$ by $\pounds_X Y^h$, $\pounds_X \omega_i$, $\pounds_X S_{ji}{}^h$, $\pounds_X \Gamma^h_{ji}$ for simplicity.

We can also prove the following formulas:

$$\pounds_X(\nabla_Y Z) - \nabla_Y(\pounds_X Z) - \nabla_{[X,Y]} Z = (\pounds_X \nabla)(Y, Z), \tag{5.13}$$

$$(\pounds_X(\nabla_Y \omega) - \nabla_Y(\pounds_X \omega) - \nabla_{[X,Y]}\omega)(Z) = -\omega((\pounds_X \nabla)(Y, Z)), \tag{5.14}$$

$$\begin{aligned}(\pounds_X(\nabla_Y S) - \nabla_Y(\pounds_X S) - \nabla_{[X,Y]} S)(U, V) = {} & (\pounds_X \nabla)(Y, S(U, V)) \\ & -S((\pounds_X \nabla)(Y, U), V) \\ & -S(U, (\pounds_X \nabla)(Y, V)),\end{aligned} \tag{5.15}$$

$$(\nabla_Y(\pounds_X \nabla))(U, V) - (\nabla_U(\pounds_X \nabla))(Y, V) = (\pounds_X K)(Y, U, V), \tag{5.16}$$

for any vector fields X, Y, Z, U, V; 1-form ω; and tensor field S of type (1, 2), or, in local coordinates,

$$\pounds_X(\nabla_j Z^h) - \nabla_j(\pounds_X Z^h) = (\pounds_X \Gamma^h_{ji}) Z^i, \tag{5.17}$$

$$\pounds_X(\nabla_j \omega_i) - \nabla_j(\pounds_X \omega_i) = -(\pounds_X \Gamma^h_{ji})\omega_h, \tag{5.18}$$

$$\pounds_X(\nabla_k S_{ji}{}^h) - \nabla_k(\pounds_X S_{ji}{}^h) = (\pounds_X \Gamma^h_{kt}) S_{ji}{}^t - (\pounds_X \Gamma^t_{kj}) S_{ti}{}^h - (\pounds_X \Gamma^t_{ki}) S_{jt}{}^h, \tag{5.19}$$

$$\nabla_k(\pounds_X \Gamma^h_{ji}) - \nabla_j(\pounds_X \Gamma^h_{ki}) = \pounds_X K_{kji}{}^h, \tag{5.20}$$

where X^h, Y^h, Z^h, ω_i, $S_{ji}{}^h$, and Γ^h_{ji} are the local components of X, Y, Z, ω, S, and ∇, respectively.

For a Riemannian manifold with metric tensor g, we have

$$\pounds_X \left\{ {h \atop ji} \right\} = \tfrac{1}{2} g^{ht}(\nabla_j(\pounds_X g_{it}) + \nabla_i(\pounds_X g_{jt}) - \nabla_t(\pounds_X g_{ji})). \tag{5.21}$$

If a point transformation in a Riemannian manifold does not change the length of any curve in the manifold, we say that the transformation does not change the metric and call the transformation a *motion* or an *isometry*.

In order that a 1-parameter group of point transformations generated by a vector field $X = X^i \partial_i$ be a group of motions, it is necessary and sufficient that X satisfy

$$\pounds_X g = 0 \quad \text{or} \quad \pounds_X g_{ji} = \nabla_j X_i + \nabla_i X_j = 0, \tag{5.22}$$

where the X_i are the covariant components of X. This equation is called *Killing's equation*, and a vector field satisfying Killing's equation is called a *Killing vector field.*

For a Killing vector field, we have

$$\pounds_X \nabla = 0,\ \pounds_X K = 0,\ \pounds_X(\nabla K) = 0, \ldots. \tag{5.23}$$

Suppose that an n-dimensional Riemannian manifold admits an r-parameter group of motions. Then the possible maximum value of r is $\frac{1}{2}n(n+1)$, and if an n-dimensional Riemannian manifold admits a $\frac{1}{2}n(n+1)$-parameter group of motions, then the manifold is of constant curvature.

If a point transformation in a manifold M with symmetric affine connection ∇ carries an arbitrary vector field parallel along an arbitrary curve into a vector field parallel along the transformed curve, we say that the transformation does not change the connection and call the transformation an *affine motion* or an *affine collineation.* In order that a 1-parameter group of transformations generated by a vector field $X = X^i\,\partial_i$ be a group of affine motions, it is necessary and sufficient that X satisfy

$$\pounds_X \nabla = 0 \quad \text{or} \quad \pounds_X \Gamma^h_{ji} = \nabla_j \nabla_i X^h + K_{kji}{}^h X^k = 0. \tag{5.24}$$

A vector field satisfying this equation is called an *affine Killing vector field,* or sometimes simply an *affine vector field.* For an affine vector field, we have

$$\pounds_X K = 0,\ \pounds_X(\nabla K) = 0, \ldots. \tag{5.25}$$

Suppose that an n-dimensional manifold M with symmetric affine connection ∇ admits an r-parameter group of affine motions. The possible maximum value of r is $n^2 + n$, and if an n-dimensional manifold with symmetric affine connection admits an $(n^2 + n)$-parameter group of affine motions, then the manifold is affinely flat, that is, the curvature tensor K vanishes.

If a point transformation in a manifold with symmetric affine connection does not change the system of paths, that is, the system of autoparallel curves, the transformation is called a *projective motion* or a *projective collineation.* In order that a 1-parameter group of transformations generated by a vector field $X = X^i\,\partial_i$ be a group of projective motions, it is necessary and sufficient that X satisfy

$$(\pounds_X \nabla)(Y, Z) = \pi(Y)Z + \pi(Z)Y \tag{5.26}$$

for any Y and Z, π being a certain 1-form, or

$$\pounds_X \Gamma^h_{ji} = \delta^h_j p_i + \delta^h_i p_j, \tag{5.27}$$

the p_i being the local components of π. A vector field satisfying this equation is called a *projective Killing vector field.* A projective Killing vector field is sometimes called simply a *projective vector field.* For a projective vector field, we have

$$(\pounds_X K)(X, Y, Z) = -(\nabla\pi)(Y, Z)X + (\nabla\pi)(X, Z)Y + (d\pi)(X, Y)Z \tag{5.28}$$

or, in local coordinates,

$$\pounds_X K_{kji}{}^h = -\delta^h_k \nabla_j p_i + \delta^h_j \nabla_k p_i + (\nabla_k p_j - \nabla_j p_k)\delta^h_i, \tag{5.29}$$

from which we have

$$\pounds_X P = 0, \tag{5.30}$$

$$(\pounds_X Q)(Y, Z, U) = -\pi(P(Y, Z)U), \tag{5.31}$$

$$\cdots$$

or

$$\pounds_X P_{kji}{}^h = 0, \tag{5.32}$$

$$\pounds_X P_{kji} = -P_{kji}{}^h p_h, \tag{5.33}$$

$$\cdots$$

in local coordinates.

Suppose that an n-dimensional manifold M with symmetric affine connection ∇ admits an r-parameter group of projective motions. Then the possible maximum value of r is $n^2 + 2n$, and if an n-dimensional manifold M with symmetric affine connection admits an $(n^2 + 2n)$-parameter group of projective motions, the manifold is projectively flat.

If a point transformation in a Riemannian manifold does not change the angle between two arbitrary vectors of the manifold, the transformation is called a *conformal motion* or a *conformal transformation.* In order that a 1-parameter group of transformations generated by a vector field $X = X^i \partial_i$ be a group of conformal transformations, it is necessary and sufficient that X satisfy

$$\pounds_X g = 2\rho g \quad \text{or} \quad \pounds_X g_{ji} = \nabla_j X_i + \nabla_i X_j = 2\rho g_{ji}, \tag{5.34}$$

ρ being a function. A vector field satisfying this equation is called a *conformal Killing vector field.* A conformal Killing vector field is sometimes called simply a *conformal vector field.* For a conformal vector field, we have

$$(\pounds_X\nabla)(Y,Z) = (d\rho)(X)Y + (d\rho)(Y)X - g(Y,Z)\,D\rho, \tag{5.35}$$

$$\begin{aligned}(\pounds_X K)(Y,Z,U) = &-(\nabla\,d\rho)(Z,U)Y + (\nabla\,d\rho)(Y,U)Z\\ &-g(Z,U)(\nabla\,D\rho)Y + g(Y,U)(\nabla\,D\rho)Z,\end{aligned} \tag{5.36}$$

$$(\pounds_X R)(Y,Z) = -(n-2)(\nabla\,d\rho)(Y,Z) + \Delta\rho\cdot g(Y,Z), \tag{5.37}$$

$$\pounds_X r = 2(n-1)\,\Delta\rho - 2\rho r, \tag{5.38}$$

where $D\rho$ is the vector field associated with the 1-form $d\rho$, or, in local coordinates,

$$\pounds_X\left\{ {h \atop ji} \right\} = \delta_j^h\rho_i + \delta_i^h\rho_j - g_{ji}\rho^h \qquad (\rho_i = \nabla_i\rho,\quad \rho^h = \rho_i g^{ih}) \tag{5.39}$$

$$\pounds_X K_{kji}{}^h = -\delta_k^h\nabla_j\rho_i + \delta_j^h\nabla_k\rho_i - \nabla_k\rho^h g_{ji} + \nabla_j\rho^h g_{ki}, \tag{5.40}$$

$$\pounds_X K_{ji} = -\,(n-2)\nabla_j\rho_i + g_{ji}\cdot\Delta\rho, \tag{5.41}$$

$$\pounds_X r = -\,2\rho r + 2(n-1)\cdot\Delta\rho, \tag{5.42}$$

from which we have

$$\pounds_X C = 0, \tag{5.43}$$

$$(\pounds_X\,D)(Y,Z,U) = -(d\rho)(C(Y,Z)U), \tag{5.44}$$

$$\cdots$$

$$\pounds_X C_{kji}{}^h = 0, \tag{5.45}$$

$$\pounds_X C_{kji} = -C_{kji}{}^h\rho_h, \tag{5.46}$$

$$\cdots$$

Suppose that an n-dimensional Riemannian manifold M admits an r-parameter group of conformal transformations. Then the possible maximum value of r is $\frac{1}{2}(n+1)(n+2)$, and if an n-dimensional Riemannian manifold admits a $\frac{1}{2}(n+1)(n+2)$-parameter group of conformal transformations, then the manifold is conformally flat.

§6. Complete Riemannian Manifolds

Suppose that there are given two points P_1 and P_2 in a connected Riemannian manifold M. We define $d(P_1, P_2)$ as the greatest lower bound of the lengths of all piecewise differentiable curves joining P_1 and

P_2. It can be verified that

$$d(P_1, P_2) \geq 0, \tag{6.1}$$

the equality holding if and only if P_1 coincides with P_2,

$$d(P_1, P_2) = d(P_2, P_1), \tag{6.2}$$

and

$$d(P_1, P_2) + d(P_2, P_3) \geq d(P_1, P_3). \tag{6.3}$$

Thus d is a metric on M. If this d is complete, that is, all Cauchy sequences converge, we say that the Riemannian metric g, or the Riemannian manifold, is *complete*. It is well-known that the following conditions on M are equivalent.

1. All Cauchy sequences converge.
2. All bounded closed sets are compact.
3. Any geodesic arc can be extended in two directions indefinitely with respect to the arc length.

It is also well-known that any points P_1 and P_2 in M can be joined by a geodesic arc whose length is equal to the distance $d(P_1, P_2)$ between these points (Hopf and Rinow [1], de Rham [3]).

When a differentiable manifold M is endowed with a Riemannian metric g, we denote this Riemannian manifold by (M, g). Let (M, g) and (M', g') be two Riemannian manifolds. If there exists a one-to-one differentiable mapping $(M, g) \rightarrow (M', g')$ such that the length of any arc in M measured by g is always equal to the length of the corresponding arc in M' measured by g', we say that the Riemannian manifold (M, g) is *isometric* with (M', g'). If there exists a one-to-one differentiable mapping $(M, g) \rightarrow (M', g')$ such that the angle of any two vectors at a point of M is always equal to that of the corresponding two vectors at the corresponding point of M', we say that the Riemannian manifold (M, g) is *conformal* to (M', g').

T. Nagano and the present author proved

THEOREM 6.1 (Yano and Nagano [1]). *If M is a complete Einstein manifold of dimension $n > 2$ admitting an infinitesimal nonisometric conformal transformation, then M is isometric to a sphere in an $(n + 1)$-dimensional Euclidean space.*

An Einstein manifold M of dimension $n > 2$ satisfies

$$R = \frac{1}{n} rg$$

with $r = \text{const.}$

On the other hand, for an infinitesimal conformal transformation X in M, we have

$$(\pounds_X R)(Y, Z) = -(n-2)(\nabla\, d\rho)(Y, Z) + \Delta\rho \cdot g(Y, Z)$$

and

$$\pounds_X r = 2(n-1)\,\Delta\rho - 2\rho r.$$

Thus if an Einstein manifold M of dimension $n > 2$ admits an infinitesimal nonisometric conformal transformation, we have, from $\pounds_X r = 0$,

$$\Delta\rho = \frac{1}{n-1}\, r\rho,$$

and consequently, from

$$\pounds_X R = \pounds_X\left(\frac{1}{n}\, r \cdot g\right) = \frac{1}{n}\, r(\pounds_X g),$$

the relation

$$\frac{1}{n}\, r(\pounds_X g)(Y, Z) = -(n-2)(\nabla\, d\rho)(Y, Z) + \frac{1}{n-1}\, r\rho g(Y, Z),$$

that is,

$$\nabla\nabla\rho = -\frac{1}{n(n-1)}\, r\rho g,$$

where ρ is not a constant function.

Generalizing the theorem above, Obata proved

THEOREM 6.2 (Obata [1, 3, 4]). *Let M be a complete Riemannian manifold of dimension $n \geq 2$. In order for M to admit a nontrivial solution ρ for the system of partial differential equations*

$$\nabla\nabla\rho = -c^2\rho g, \qquad c = \text{const} > 0,$$

it is necessary and sufficient that M be isometric to a sphere S of radius $1/c$ in the $(n+1)$-dimensional Euclidean space.

Ishihara and Tashiro proved

THEOREM 6.3 (Ishihara and Tashiro [1], Tashiro [1]). *Let M be a complete Riemannian manifold of dimension $n \geq 2$. In order for M to admit a nontrivial solution ρ for the system of partial differential equations*

$$\nabla\nabla\rho = -\frac{1}{n}\Delta\rho \cdot g,$$

it is necessary and sufficient that M be conformal to a sphere in the $(n+1)$-*dimensional Euclidean space.*

In the sequel we mainly study compact Riemannian manifolds. Since a compact Riemannian manifold is complete, these theorems are also valid for compact Riemannian manifolds and so will prove very useful in the later discussions.

PROBLEMS

1. Prove that

$$**\omega = (-1)^{p(n+1)}\omega,$$

where ω is a p-form, and that

$$\langle *\omega, *\theta\rangle = \langle \omega, \theta\rangle$$

where ω and θ are both p-forms.

2. Prove that Euler's differential equations of the integral

$$\int_{t_1}^{t_2} \sqrt{g_{ji}(x(t))\frac{dx^j}{dt}\frac{dx^i}{dt}}\,dt$$

are

$$\frac{d^2x^h}{ds^2} + \begin{Bmatrix} h \\ ji \end{Bmatrix} \frac{dx^j}{ds}\frac{dx^i}{ds} = 0,$$

s being the arc length, that is, a parameter satisfying

$$g_{ji}\frac{dx^j}{ds}\frac{dx^i}{ds} = 1.$$

3. Prove that, in a nonempty intersection of two coordinate neighborhoods $\{U; x^h\}$ and $\{U'; x^{h'}\}$, the Christoffel symbols $\begin{Bmatrix} h \\ ji \end{Bmatrix}$ and $\begin{Bmatrix} h' \\ j'i' \end{Bmatrix}$ are related by

$$\frac{\partial x^h}{\partial x^{h'}}\begin{Bmatrix} h' \\ j'i' \end{Bmatrix} = \frac{\partial x^j}{\partial x^{j'}}\frac{\partial x^i}{\partial x^{i'}}\begin{Bmatrix} h \\ ji \end{Bmatrix} + \frac{\partial^2 x^h}{\partial x^{j'}\,\partial x^{i'}}.$$

4. The covariant derivative $\nabla_X S$ along a vector field X of a tensor field S of type $(1, 2)$ being defined by

$$(\nabla_X S)(Y, Z) = \nabla_X(S(Y, Z)) - S(\nabla_X Y, Z) - S(Y, \nabla_X Z),$$

prove that $\nabla_X S$ has local components

$$X^k\nabla_k S_{ji}{}^h = X^k\left(\partial_k S_{ji}{}^h + \begin{Bmatrix} h \\ kt \end{Bmatrix} S_{ji}{}^t - \begin{Bmatrix} t \\ kj \end{Bmatrix} S_{ti}{}^h - \begin{Bmatrix} t \\ ki \end{Bmatrix} S_{jt}{}^h\right),$$

where $S_{ji}{}^h$, X^h, Y^h, Z^h are local components of S, X, Y, Z, respectively; that is,

$$S(Y, Z) = (S_{ji}{}^h Y^j X^i)\,\partial_h, \quad X = X^i\,\partial_i, \quad Y = Y^i\,\partial_i, \quad Z = Z^i\,\partial_i.$$

5. Prove that

$$\nabla g = 0, \qquad \nabla I = 0, \qquad \nabla e = 0,$$

∇ denoting covariant differentiation with respect to the Christoffel symbols, that is, with respect to the Riemannian connection.

6. Prove that

$$d(d\omega) = 0, \qquad \delta(\delta\omega) = 0,$$

for any p-form ω.

7. The covariant derivative $\nabla_X Y$ of a vector field Y along a vector field X has the property

$$\nabla_{fX}(gY) = f\{(\nabla_X g)Y + g(\nabla_X Y)\}.$$

for any functions f and g. Prove that the expression

$$\nabla_X \nabla_Y Z - \nabla_Y \nabla_X Z - \nabla_{[X,Y]} Z$$

defines a tensor field of type (1, 3), that is,

$$\nabla_{fX}\nabla_{gY}(hZ) - \nabla_{gY}\nabla_{fX}(hZ) - \nabla_{[fX,gY]}(hZ) = fgh(\nabla_X\nabla_Y Z - \nabla_Y\nabla_X Z - \nabla_{[X,Y]}Z),$$

for any functions f, g, h.

8. Putting

$$\nabla_X \nabla_Y Z - \nabla_Y \nabla_X Z - \nabla_{[X,Y]} Z = K(X, Y)Z$$

and denoting by X^h, Y^h, Z^h, and $K_{kji}{}^h$ the local components of X, Y, Z, and K, respectively, where

$$X = X^i\,\partial_i, \qquad Y = Y^i\,\partial_i, \qquad Z = Z^i\,\partial_i,$$

$$K(X, Y)Z = (K_{kji}{}^h X^k Y^j Z^i)\,\partial_h,$$

prove that

$$\nabla_k \nabla_j Z^h - \nabla_j \nabla_k Z^h = K_{kji}{}^h Z^i.$$

9. Prove that the vanishing of all the Christoffel symbols is equivalent to the fact that all the components of the metric tensor g are constants, and use this to prove: A necessary and sufficient condition for a Riemannian manifold M to be covered by a system of coordinate neighborhoods $\{U; x^h\}$ such that the components of the fundamental metric tensor are all constant in any coordinate neighborhood is that the Riemann-Christoffel curvature tensor vanishes identically.

10. Prove that

$$\nabla_k \nabla_j \omega_i - \nabla_j \nabla_k \omega_i = -K_{kji}{}^h \omega_h$$

for an arbitrary 1-form ω and that

$$\nabla_l \nabla_k S_{ji}{}^h - \nabla_k \nabla_l S_{ji}{}^h = K_{lkt}{}^h S_{ji}{}^t - K_{lkj}{}^t S_{ti}{}^h - K_{lki}{}^t S_{jt}{}^h,$$

for an arbitrary tensor field S of type (1, 2).

11. (i) Use

$$\nabla_X \nabla_Y (\nabla_Z f) - \nabla_Y \nabla_X (\nabla_Z f) - \nabla_{[X,Y]}(\nabla_Z f) = -(\nabla f)(K(X, Y)Z)$$

and

$$\nabla_X(\nabla_Y \nabla_Z f - \nabla_Z \nabla_Y f - \nabla_{[Y,Z]} f) = 0$$

for an arbitrary function f to prove

$$K(X, Y)Z + K(Y, Z)X + K(Z, X)Y = 0$$

for arbitrary vector fields X, Y, Z.

(ii) Use

$$(\nabla_X \nabla_Y g - \nabla_Y \nabla_X g - \nabla_{[X,Y]} g)(U, V) = -g(K(X, Y)U, V) - g(U, K(X, Y)V)$$

to prove

$$K(X, Y; U, V) + K(X, Y; V, U) = 0.$$

(iii) Use

$$K(X, Y)Z + K(Y, Z)X + K(Z, X)Y = 0$$

and

$$K(X, Y; U, V) + K(X, Y; V, U) = 0$$

for arbitrary vector fields X, Y, Z, U to prove

$$R(X, Y) = R(Y, X).$$

(iv) Use

$$K(X, Y; U, V) + K(Y, X; U, V) = 0,$$

$$K(X, Y; U, V) + K(Y, U; X, V) + K(U, X; Y, V) = 0,$$

and

$$K(X, Y; U, V) + K(X, Y; V, U) = 0$$

to prove

$$K(X, Y; U, V) = K(U, V; X, Y).$$

[Hint:
$$\begin{aligned} K(X, Y; U, V) &= -K(Y, U; X, V) - K(U, X; Y, V) \\ &= +K(Y, U; V, X) + K(U, X; V, Y) \\ &= -K(U, V; Y, X) - K(V, Y; U, X) \\ &\quad -K(X, V; U, Y) - K(V, U; X, Y) \\ &= 2K(U, V; X, Y) + \{K(V, Y; X, U) + K(X, V; Y, U)\} \\ &= 2K(U, V; X, Y) - K(Y, X; V, U) \\ &= 2K(U, V; X, Y) - K(X, Y; U, V).] \end{aligned}$$

(v) Use

$$\nabla_X \nabla_Y (\nabla_Z U) - \nabla_Y \nabla_X (\nabla_Z U) - \nabla_{[X,Y]} (\nabla_Z U) = K(X, Y)(\nabla_Z U)$$

and

$$\nabla_Z (\nabla_X \nabla_Y U - \nabla_Y \nabla_X U - \nabla_{[X,Y]} U) = \nabla_Z (K(X, Y)U)$$

for arbitrary vector fields X, Y, Z, U to prove the Bianchi identity

$$(\nabla_X K)(Y, Z) + (\nabla_Y K)(Z, X) + (\nabla_Z K)(X, Y) = 0.$$

12. The affine connection ∇ has the property:

$$\nabla_{fX}(gY) = f\{(\nabla_X g)Y + g\nabla_X Y\}.$$

Prove that the difference of two affine connections ∇ and ∇^* defines a tensor field of type (1, 3):

$$H(X, Y) = (\nabla_X^* - \nabla_X)Y.$$

13. (i) Prove that if two affine connections ∇ and ∇^*, both without torsion, give the same systems of paths, then the difference H of the two connections is of the form

$$H(X, Y) = \pi(X)Y + \pi(Y)X,$$

π being a certain 1-form, or

$$\Gamma^{*h}_{ji} - \Gamma^h_{ji} = \delta^h_j p_i + \delta^h_i p_j$$

in local coordinates, Γ^{*h}_{ji}, Γ^h_{ji}, p_i being, respectively, local components of ∇^*, ∇, π (Eisenhart [2], Weyl [1, 3]).

(ii) Suppose that two affine connections ∇ and ∇^*, both without torsion, are projectively related as in (i). Denote by K and K^* the curvature tensors of ∇ and ∇^*, respectively. Prove that

$$K^*(X, Y)Z = K(X, Y)Z - p(Y, Z)X + p(X, Z)Y + \{p(X, Y) - p(Y, X)\}Z,$$

where

$$p(X, Y) = (\nabla_X \pi)(Y) - \pi(X)\pi(Y),$$

or

$$K^{*}_{kji}{}^h = K_{kji}{}^h - \delta^h_k p_{ji} + \delta^h_j p_{ki} + (p_{kj} - p_{jk})\delta^h_i,$$

where

$$p_{ji} = \nabla_j p_i - p_j p_i$$

in local coordinates.

(iii) We define

$$L(X, Y) = \frac{1}{n^2 - 1}\{-nR(X, Y) + R(Y, X)\},$$

and $L^*(X, Y)$ by a similar expression.

Use the result in (ii) to prove

$$L^* = L + p$$

or

$$P^*_{ji} = P_{ji} + p_{ji}$$

in local coordinates, where P^*_{ji}, P_{ji}, p_{ji} are components of L^*, L, p, respectively.

(iv) Use the results of (ii) and (iii) to prove

$$P^*(X, Y)Z = P(X, Y)Z$$

where

$$P(X, Y)Z = K(X, Y)Z + L(Y, Z)X - L(X, Z)Y - \{L(X, Y) - L(Y, X)\}Z,$$

$P^*(X, Y)Z$ having a similar expression, or

$$P^{*}_{kji}{}^h = P_{kji}{}^h,$$

where

$$P_{kji}{}^h = K_{kji}{}^h + \delta^h_k P_{ji} - \delta^h_j P_{ki} - (P_{kj} - P_{jk})\,\delta^h_i$$

in local coordinates, $P^{*}_{kji}{}^h$ having a similar expression.

(v) Prove that

$$Q^*(X, Y, Z) = Q(X, Y, Z) + \pi(P(X, Y)Z),$$

where

$$Q(X, Y, Z) = (\nabla_X L)(Y, Z) - (\nabla_Y L)(X, Z),$$

$Q^*(X, Y, Z)$ having a similar definition, or

$$P^*_{kji} = P_{kji} + P_{kji}{}^h p_h,$$

where

$$P_{kji} = \nabla_k P_{ji} - \nabla_j P_{ki}$$

in local coordinates, P^*_{kji} having a similar expression.

(vi) Prove the theorem of Weyl: A necessary and sufficient condition for a manifold with a symmetric affine connection ∇ to be projectively flat is that

$$P = 0 \quad \text{for} \quad n > 2$$

and

$$Q = 0 \quad \text{for} \quad n = 2.$$

(vii) Prove that a projectively flat Riemannian manifold is of constant curvature.

14. (i) Suppose that two Riemannian metrics g and g^* are conformally related, $g^* = \rho^2 g$, and denote by ∇ and ∇^* the Riemannian connections defined by g and g^*, respectively. Prove that

$$(\nabla^* - \nabla)(X, Y) = \omega(X)Y + \omega(Y)X - g(X, Y)U$$

for any vector fields X, Y, where

$$\omega = d \log \rho$$

and U is a vector field defined by

$$g(U, X) = \omega(X),$$

or

$$\begin{Bmatrix} h \\ ji \end{Bmatrix}^* - \begin{Bmatrix} h \\ ji \end{Bmatrix} = \delta^h_j \rho_i + \delta^h_i \rho_j - g_{ji}\rho^h$$

in local coordinates, where

$$\rho_i = \nabla_i \log \rho, \qquad \rho^h = \rho_i g^{ih}.$$

(ii) Suppose that two Riemannian metrics g and g^* are conformally related, $g^* = \rho^2 g$, and denote by K and K^* the curvature tensors of g and g^*, respectively. Prove that

$$K^*(X, Y)Z = K(X, Y)Z - s(Y, Z)X + s(X, Z)Y - g(Y, Z)SX + g(X, Z)SY$$

for any vector fields X, Y, Z, where s is a tensor field of type (0, 2) defined by

$$s(X, Y) = (\nabla_X \omega)(Y) - \omega(X)\omega(Y) + \tfrac{1}{2}\omega(U)g(X, Y)$$

and S is a tensor field of type (1, 1) defined by

$$g(SX, Y) = s(X, Y),$$

or

$$K^*_{kji}{}^h = K_{kji}{}^h - \delta^h_k \rho_{ji} + \delta^h_j \rho_{ki} - \rho_k{}^h g_{ji} + \rho_j{}^h g_{ki},$$

where

$$\rho_{ji} = \nabla_j \rho_i - \rho_j \rho_i + \tfrac{1}{2}\rho_t \rho^t g_{ji},$$
$$\rho_k{}^h = \rho_{kt} g^{th}$$

in local coordinates.

(iii) We define

$$L(X, Y) = -\frac{1}{n-2} R(X, Y) + \frac{1}{2(n-1)(n-2)} r g(X, Y),$$

and L^* by a similar expression. Use the results of (ii) to prove

$$L^* = L + s,$$

or

$$C^*_{ji} = C_{ji} + \rho_{ji},$$

where

$$C_{ji} = -\frac{1}{n-2} K_{ji} + \frac{1}{2(n-1)(n-2)} r g_{ji},$$

C^*_{ji} having a similar expression.

(iv) Use the results of (ii) and (iii) to prove

$$C^* = C,$$

where

$$C(X, Y)Z = K(X, Y)Z + L(Y, Z)X - L(X, Z)Y + g(Y, Z)NX - g(X, Z)NY,$$

and

$$g(NX, Y) = L(X, Y)$$

for any vector fields X, Y, and Z, C^* having a similar expression, or

$$C^*_{kji}{}^h = C_{kji}{}^h,$$

where

$$C_{kji}{}^h = K_{kji}{}^h + \delta^h_k C_{ji} - \delta^h_j C_{ki} + C_k{}^h g_{ji} - C_j{}^h g_{ki},$$
$$C_k{}^h = C_{kt} g^{th}$$

in local coordinates, and $C^*_{kji}{}^h$ has a similar expression.

(v) Prove that

$$D^*(X, Y, Z) = D(X, Y, Z) + \omega(C(X, Y)Z),$$

where

$$D(X, Y, Z) = (\nabla_X L)(Y, Z) - (\nabla_Y L)(X, Z),$$

$D^*(X, Y, Z)$ having a similar expression, or

$$C^*_{kji} = C_{kji} + C_{kji}{}^h \rho_h$$

in local coordinates, where

$$C_{kji} = \nabla_k C_{ji} - \nabla_j C_{ki},$$

C^*_{kji} having a similar expression.

(vi) Prove the theorem of Weyl: A necessary and sufficient condition for a Riemannian manifold to be conformally flat is that

$$C = 0 \quad \text{for} \quad n > 3$$

and

$$D = 0 \quad \text{for} \quad n = 3.$$

15. Let Z, ω, and S be, respectively, a vector field, a 1-form and a tensor field of type (1, 2). Prove the formulas

$$£_X(\nabla_Y Z) - \nabla_Y(£_X Z) - \nabla_{[X,Y]} Z = (£_X \nabla)(Y, Z),$$
$$(£_X(\nabla_Y \omega) - \nabla_Y(£_X \omega) - \nabla_{[X,Y]}\omega)(U) = -\omega((£_X \nabla)(Y, U)),$$
$$(£_X(\nabla_Y S) - \nabla_Y(£_X S) - \nabla_{[X,Y]} S)(U, V) = (£_X \nabla)(Y, S(U, V))$$
$$-S((£_X \nabla)(Y, U), V)$$
$$-S(U, (£_X \nabla)(Y, V))$$

for any vector fields X, Y, U, V, or

$$\pounds_X(\nabla_j Z^h) - \nabla_j(\pounds_X Z^h) = (\pounds_X \Gamma^h_{ji})Z^i,$$
$$\pounds_X(\nabla_j \omega_i) - \nabla_j(\pounds_X \omega_i) = -(\pounds_X \Gamma^h_{ji})\omega_h,$$
$$\pounds_X(\nabla_k S_{ji}{}^h) - \nabla_k(\pounds_X S_{ji}{}^h) = (\pounds_X \Gamma^h_{kt})S_{ji}{}^t - (\pounds_X \Gamma^t_{kj})S_{ti}{}^h - (\pounds_X \Gamma^t_{ki})S_{jt}{}^h$$

in local coordinates.

16. Prove that

$$(\nabla_Y(\pounds_X \nabla))(U, V) - (\nabla_U(\pounds_X \nabla))(Y, V) = (\pounds_X K)(Y, U, V)$$

for any vector fields X, Y, U, V, or

$$\nabla_k(\pounds_X \Gamma^h_{ji}) - \nabla_j(\pounds_X \Gamma^h_{ki}) = \pounds_X K_{kji}{}^h$$

in local coordinates.

17. Prove: In a Riemannian manifold with the metric tensor g,

$$\pounds_X \begin{Bmatrix} h \\ ji \end{Bmatrix} = \tfrac{1}{2} g^{ht}(\nabla_j(\pounds_X g_{it}) + \nabla_i(\pounds_X g_{jt}) - \nabla_t(\pounds_X g_{ji})).$$

18. Prove: A Killing vector field X in a Riemannian manifold has the properties

$$\pounds_X \nabla = 0,$$
$$\pounds_X K = 0,$$
$$\pounds_X(\nabla K) = 0,$$
$$\cdots.$$

19. A Killing vector field $X = X^i\,\partial_i$ in a Riemannian manifold satisfies

$$\pounds_X g_{ji} = \nabla_j X_i + \nabla_i X_j = 0,$$

where $X_i = g_{ih}X^h$, from which we have

$$\pounds_X \begin{Bmatrix} h \\ ji \end{Bmatrix} = \nabla_j \nabla_i X^h + K_{kji}{}^h X^k = 0.$$

Thus, putting

$$X_i{}^h = \nabla_i X^h,$$

we have a system of partial differential equations with unknown functions X^h and $X_i{}^h$:

$$\begin{cases} \nabla_i X^h = X_i{}^h, \\ \nabla_j X_i{}^h = -K_{kji}{}^h X^k \end{cases}$$

with conditions

$$X_{ji} + X_{ij} = 0 \qquad (X_{ji} = X_j{}^h g_{hi}).$$

Prove that a necessary and sufficient condition for this system of partial differential equations to be completely integrable is that the Riemannian manifold be of constant curvature and that the solution contain $\frac{1}{2}n(n + 1)$ arbitrary constants.

20. Prove: An affine Killing vector field X in a manifold with affine connection ∇ has the properties

$$\pounds_X K = 0,$$
$$\pounds_X(\nabla K) = 0,$$
$$\cdots.$$

21. An affine Killing vector field $X = X^i \partial_i$ in a manifold with affine connection ∇ satisfies

$$£_X \Gamma^h_{ji} = \nabla_j \nabla_i X^h + K_{kji}{}^h X^k = 0.$$

Thus, putting

$$X_i{}^h = \nabla_i X^h,$$

we have a system of partial differential equations with unknown functions X^h and $X_i{}^h$:

$$\begin{cases} \nabla_i X^h = X_i{}^h, \\ \nabla_j X_i{}^h = -K_{kji}{}^h X^k. \end{cases}$$

Prove: A necessary and sufficient condition for this system of partial differential equations to be completely integrable is that the manifold with this affine connection be affinely flat and that the solution contain $n^2 + n$ arbitrary constants.

22. Prove: A projective Killing vector field X in a manifold with a symmetric affine connection ∇ has the properties

$$£_X P = 0,$$
$$(£_X s)(Y, Z, U) = -\omega(P(Y, Z)U),$$
$$\cdots.$$

23. A projective Killing vector field $X = X^i \partial_i$ in a manifold with a symmetric affine connection ∇ satisfies

$$£_X \Gamma^h_{ji} = \nabla_j \nabla_i X^h + K_{kji}{}^h X^k = \delta^h_j p_i + \delta^h_i p_j$$

for a certain 1-form p_i.

Hence

$$£_X K_{kji}{}^h = -\delta^h_k \nabla_j p_i + \delta^h_j \nabla_k p_i + (\nabla_k p_j - \nabla_j p_k)\, \delta^h_i,$$

and consequently

$$\nabla_j p_i = £_X P_{ji},$$

where

$$P_{ji} = -\frac{1}{n^2 - 1}(nK_{ji} + K_{ij}).$$

Thus, putting

$$X_i{}^h = \nabla_i X^h,$$

we have a system of partial differential equations with unknown functions X^h, $X_i{}^h$, and p_i:

$$\begin{cases} \nabla_i X^h = X_i{}^h, \\ \nabla_j X_i{}^h = -K_{kji}{}^h X^k + \delta^h_j p_i + \delta^h_i p_j, \\ \nabla_j p_i = X^k \nabla_k L_{ji} + X_j{}^t P_{ti} + X_i{}^t P_{jt}. \end{cases}$$

Prove: A necessary and sufficient condition for this system of partial differential equations to be completely integrable is that the manifold with this affine connection be projectively flat and that the solution contain $n^2 + 2n$ arbitrary constants.

24. Prove: A conformal Killing vector field X in a Riemannian manifold with the metric tensor g, that is, a vector field X satisfying $£_X g = 2\rho g$; has properties

$$£_X C = 0,$$
$$(£_X D)(Y, Z, U) = -(d\rho)(C(Y, Z)U),$$
$$\cdots.$$

25. A conformal Killing vector field $X = X^i \partial_i$ in a Riemannian manifold with metric tensor g satisfies

$$£_X g_{ji} = \nabla_j X_i + \nabla_i X_j = 2\rho g_{ji}$$

for a particular function ρ. Hence

$$£_X \left\{ {h \atop ji} \right\} = \delta_j^h \rho_i + \delta_i^h \rho_j - g_{ji}\rho^h,$$

where $\rho_i = \nabla_i \rho$ and $\rho^h = \rho_i g^{ih}$ and

$$£_X K_{kji}{}^h = -\delta_k^h \nabla_j \rho_i + \delta_j^h \nabla_k \rho_i - \nabla_k \rho^h g_{ji} + \nabla_j \rho^h g_{ki},$$

and consequently

$$\nabla_j \rho_i = £_X C_{ji},$$

where

$$C_{ji} = -\frac{1}{n-2} K_{ji} + \frac{1}{2(n-1)(n-2)} r g_{ji}.$$

Thus, putting

$$X_i{}^h = \nabla_i X^h,$$

we have a system of partial differential equations with unknown functions X^h, $X_i{}^h$; ρ, ρ_i:

$$\begin{cases} \nabla_i X^h = X_i{}^h, \\ \nabla_j X_i{}^h = -K_{kji}{}^h X^k + \delta_j^h \rho_i + \delta_i^h \rho_j - g_{ji}\rho^h, \\ \nabla_i \rho = \rho_i, \\ \nabla_j \rho_i = X^k \nabla_k C_{ji} + X_j{}^t C_{ti} + X^h C_{jt} \end{cases}$$

with the conditions

$$X_{ji} + X_{ij} = 2\rho g_{ji} \quad (X_{ji} = X_j{}^h g_{hi}).$$

Prove: A necessary and sufficient condition for this system of partial differential equations to be completely integrable is that the Riemannian manifold be conformally flat and that the solution contain $\frac{1}{2}(n+1)(n+2)$ arbitrary constants.

CHAPTER

2

Harmonic 1-Forms and Killing Vector Fields

In this chapter we use integral formulas to study properties of harmonic 1-forms, as well as Killing, affine Killing, projective Killing, and conformal Killing vector fields.

§1. Integral Formulas

In this chapter we consider compact and orientable Riemannian manifolds M without boundary and with positive-definite fundamental metric tensor g. We already have

PROPOSITION 1.1. *The integral formula*

$$\int_M \Delta f\, dV = 0 \tag{1.1}$$

holds for any function f in M, Δf being defined by

$$\Delta f = \delta\, df = -g^{ji}\, \nabla_j \nabla_i f.$$

Thus, if $\Delta f \geq 0$ (or $\Delta f \leq 0$) in M, then we must have $\Delta f = 0$. Since

$$\tfrac{1}{2}\Delta f^2 = (\Delta f)f - |\nabla f|^2,$$

we find, by Proposition 1.1,

$$\int_M [(\Delta f)f - |\nabla f|^2]\, dV = 0.$$

Thus, if $\Delta f \geq 0$ (or $\Delta f \leq 0$) in M, then $\Delta f = 0$ in M and, consequently, from the integral formula above,

$$\int_M |\nabla f|^2 \, dV = 0,$$

from which we have $\nabla f = 0$. Consequently, $f = \text{const}$ in M. Thus we have proved

PROPOSITION 1.2. *If $\Delta f \geq 0$ (or $\Delta f \leq 0$) in M, then f is constant in M.*

This proposition is sometimes called Bochner's lemma.

If a nonconstant function f in M satisfies

$$\Delta f = cf,$$

c being a constant different from zero, then

$$\int_M [cf^2 - |\nabla f|^2] \, dV = 0,$$

from which we see that c must be positive. Thus we have

PROPOSITION 1.3. *If $\Delta f = cf$ for a nonconstant function f in M, c being a constant, c must be positive.*

We next consider a vector field X in M and apply the integral formula (1.1) to $f = |X|^2$. Then

$$\int_M [(g^{ji}\nabla_j \nabla_i X^h)X_h + |\nabla X|^2] \, dV = 0, \tag{1.2}$$

from which we have

PROPOSITION 1.4. *If the second covariant derivative $\nabla\nabla X$ of a vector field X vanishes, then the first covariant derivative ∇X vanishes* (Bochner [4, 5]).

Suppose that there is given a vector field $X = X^i \partial_i$. Then with this vector field X is associated a 1-form $\xi = g_{ji}X^j dx^i = X_i \, dx^i$.

For this 1-form ξ we have

$$\begin{aligned} d\xi &: \quad \nabla_j X_i - \nabla_i X_j, \\ \delta \, d\xi &: \quad -g^{kj} \, \nabla_k \nabla_j X_i + g^{kj} \, \nabla_k \nabla_i X_j, \end{aligned}$$

and

$$\begin{aligned} \delta\xi &: \quad -g^{kj} \, \nabla_k X_j, \\ d \, \delta\xi &: \quad -g^{kj} \, \nabla_i \nabla_k X_j. \end{aligned}$$

Consequently,

$$\Delta\xi = (\delta d + d\delta)\xi\colon \quad -g^{kj}\nabla_k\nabla_j X_i + K_i{}^h X_h \tag{1.3}$$

by virtue of the relation

$$\begin{aligned} g^{kj}(\nabla_k\nabla_i X_j - \nabla_i\nabla_k X_j) &= -g^{kj}K_{kij}{}^h X_h \\ &= K_i{}^h X_h. \end{aligned}$$

We denote by ΔX the vector field which is associated with the 1-form $\Delta\xi$; thus

$$\Delta X\colon \quad -(g^{kj}\nabla_k\nabla_j X^h - K_i{}^h X^i). \tag{1.4}$$

Take a function f and form

$$\Delta f = (\delta d + d\delta)f.$$

Then

$$d\,\Delta f = d\delta\, df;$$

that is,

$$d\,\Delta f = \Delta\, df, \qquad D\,\Delta f = \Delta\, Df. \tag{1.5}$$

Take next a function f and a vector field $X = X^i\,\partial_i$ and form

$$\nabla_i(fX^i) = (\nabla_i f)X^i + f\,\nabla_i X^i.$$

Integrating this over the manifold M, we find

$$\int_M (\nabla_i f)X^i\,dV + \int_M f\nabla_i X^i\,dV = 0,$$

which can be written as

$$\int_M g(Df, X)\,dV = \int_M f(\delta X)\,dV. \tag{1.6}$$

In the sequel we shall meet the expression

$$-\,g^{kj}\nabla_k\nabla_j X^h - K_i{}^h X^i;$$

so we denote this vector field by

$$\Box X\colon \quad -\,(g^{kj}\nabla_k\nabla_j X^h + K_i{}^h X^i). \tag{1.7}$$

If we denote by Q the tensor field of type (1, 1) whose local components are $K_i{}^h$, then $\Box X$ can be represented as

$$\Box X = \Delta X - 2QX. \tag{1.8}$$

Consider a vector field $X = X^i\,\partial_i$ in M and form another vector field

$$X^j(\nabla_j X^i) - (\nabla_j X^j)X^i.$$

Applying the operator ∇_i to this, we obtain

$$\begin{aligned}\nabla_i[X^j(\nabla_j X^i) - (\nabla_j X^j)X^i] &= (\nabla_i X^j)(\nabla_j X^i) + X^j(\nabla_i\nabla_j X^i)\\ &\quad - (\nabla_i\nabla_j X^j)X^i - (\nabla_j X^j)^2\\ &= K_{ji}X^jX^i + (\nabla^j X^i)(\nabla_i X_j) - (\nabla_j X^j)^2\end{aligned}$$

by virtue of the identity

$$\nabla_i\nabla_j X^i - \nabla_j\nabla_i X^i = K_{ji}X^i.$$

Thus we have the integral formula

$$\int_M [K_{ji}X^jX^i + (\nabla^j X^i)(\nabla_i X_j) - (\nabla_j X^j)^2]\,dV = 0, \tag{1.9}$$

which can also be written as

$$\int_M [R(X, X) - \tfrac{1}{2}\,|d\xi|^2 + |\nabla X|^2 - (\delta\xi)^2]\,dV = 0, \tag{1.10}$$

or

$$\int_M [R(X, X) + \tfrac{1}{2}\,|\pounds_X g|^2 - |\nabla X|^2 - (\delta X)^2]\,dV = 0, \tag{1.11}$$

where ξ is a 1-form associated with X.

Forming $-(1.2) + (1.10)$, we find

$$\int_M [g(\Delta X, X) - \tfrac{1}{2}\,|d\xi|^2 - (\delta X)^2]\,dV = 0. \tag{1.12}$$

Forming $-(1.2) - (1.11)$, we find

$$\int_M [g(\square X, X) - \tfrac{1}{2}\,|\pounds_X g|^2 + (\delta X)^2]\,dV = 0. \tag{1.13}$$

Note that formula (1.13) can also be written as

$$\int_M \left[g\left(\square X + \frac{n-2}{n} D\,\delta X, X\right) - \frac{1}{2}\left|\pounds_X g + \frac{2}{n}(\delta X)g\right|^2\right] dV = 0. \tag{1.14}$$

§2. Harmonic 1-Forms

A vector field X, or a 1-form ξ associated with it, is harmonic if it satisfies

$$d\xi = 0, \qquad \delta\xi = 0. \tag{2.1}$$

Suppose that a harmonic 1-form ξ is the differential of a function f:

$$\xi = df;$$

then we have

$$\delta\xi = \delta\, df = \Delta f = 0,$$

from which, by Proposition 1.2, $f = \text{const}$, and consequently ξ vanishes identically. Thus we have

PROPOSITION 2.1. *If a harmonic 1-form in M is the differential of a function, then it is identically zero.*

From (1.10), we have

PROPOSITION 2.2. *For a vector field X in M,*

$$\int_M [R(X, X) + |\nabla X|^2]\, dV \geq 0, \tag{2.2}$$

equality holding if and only if X is a harmonic vector (Watanabe [1]).

Since

$$\int_M [R(X, X) + |\nabla X|^2]\, dV = 0$$

for a harmonic vector field X, we have

THEOREM 2.3. *If the Ricci curvature in M satisfies* $R(X, X) \geq 0$, *then a harmonic vector field X in M has a vanishing covariant derivative. If the Ricci curvature in M is positive-definite, then a harmonic vector field other than zero does not exist in M* (Bochner [2]).

Thus, using Hodge's theorem, we have

THEOREM 2.4. *In a compact and orientable manifold with positive Ricci curvature, the first Betti number vanishes* (Bochner [2], Myers [1]).

Suppose that X is a harmonic vector field; then $d\xi = 0$, $\delta\xi = 0$, where ξ is the 1-form associated with X. Then

$$\Delta\xi = (\delta d + d\delta)\xi = 0;$$

that is,

$$g^{kj}\nabla_k\nabla_j X^h - K_i{}^h X^i = 0. \tag{2.3}$$

Conversely, if $\Delta\xi = 0$ or (2.3) holds, then, from (1.12),

$$\int_M [\tfrac{1}{2}\, |d\xi|^2 + (\delta\xi)^2]\, dV = 0,$$

from which

$$d\xi = 0, \qquad \delta\xi = 0;$$

that is, X is a harmonic vector field. Thus,

THEOREM 2.5. *In order that a vector field X be harmonic, it is necessary and sufficient that*

$$\Delta\xi = 0,$$

or

$$g^{kj}\nabla_k\nabla_j X^h - K_i{}^h X^i = 0,$$

where ξ is the 1-form associated with X.

§3. Killing Vector Fields

A vector field X is a Killing vector field if it satisfies

$$\pounds_X g_{ji} = \nabla_j X_i + \nabla_i X_j = 0. \tag{3.1}$$

Consequently,

$$\nabla_i X^i = 0. \tag{3.2}$$

Suppose that a 1-form associated with a Killing vector field X is the differential of a function f

$$X_i = \nabla_i f;$$

then

$$\Delta f = 0,$$

from which we have $f = \text{const}$ and $X = 0$.

Thus we have

PROPOSITION 3.1. *If the 1-form associated with a Killing vector field is the differential of a function, then it is identically zero.*

From (1.11), we have

PROPOSITION 3.2. *For a vector field X in M,*

$$\int_M [R(X, X) - |\nabla X|^2 - (\delta X)^2]\, dV \leq 0, \tag{3.3}$$

equality holding if and only if X is a Killing vector field (Watanabe [1]).

Since

$$\int_M [R(X, X) - |\nabla X|^2]\, dV = 0$$

for a Killing vector field X, we have

THEOREM 3.3. *If the Ricci curvature in M satisfies $R(X, X) \leq 0$, then a Killing vector field X in M has vanishing covariant derivative. If the Ricci curvature in M is negative-definite, then a Killing vector field other than zero does not exist in M* (Bochner [2]).

Suppose that X is a Killing vector field in M, then from the identity

$$\nabla_h \nabla_j X_i - \nabla_j \nabla_h X_i = -K_{hji}{}^t X_t$$

or

$$\nabla_h \nabla_j X_i + \nabla_j \nabla_i X_h = -K_{hji}{}^t X_t,$$

we find

$$g^{kj} \nabla_k \nabla_j X_h + K_h{}^t X_t = 0, \tag{3.4}$$

or

$$g^{kj} \nabla_k \nabla_j X^h + K_i{}^h X^i = 0, \quad \text{that is,} \quad \Box X = 0. \tag{3.5}$$

Conversely, if (3.5) and $\delta\xi = 0$ are satisfied, where ξ is the 1-from associated with X, then, from (1.13),

$$\pounds_X g = 0;$$

that is, X is a Killing vector field. Thus

THEOREM 3.4. *In order that a vector field X be a Killing vector field, it is necessary and sufficient that*

$$\Box X = 0 \quad \textit{and} \quad \delta X = 0.$$

Suppose that ω_i is a harmonic 1-form and X^h is a Killing vector field. Then

$$\begin{aligned}\Delta(\omega_h X^h) &= -(g^{ji} \nabla_j \nabla_i \omega_h) X^h - (\nabla_i \omega_h)(\nabla^i X^h) - \omega_h (g^{ji} \nabla_j \nabla_i X^h) \\ &= -K_h{}^i \omega_i X^h - (\nabla_i \omega_h)(\nabla^i X^h) + \omega_h K_i{}^h X^i \\ &= 0\end{aligned}$$

by virtue of (2.3), (3.4), and the formulas $\nabla_i \omega_h = \nabla_h \omega_i$ and $\nabla^i X^h = -\nabla^h X^i$. Consequently, by Proposition 1.2,

$$\omega_h X^h = \text{const.}$$

Thus,

PROPOSITION 3.5. *For a harmonic 1-form ω and a Killing vector field X, the inner product $\omega(X)$ is constant in the manifold* (Bochner [2]).

§4. Affine Killing Vector Fields

A vector field X is an affine Killing vector field if it satisfies

$$\pounds_X \nabla = 0, \quad \text{that is,} \quad \nabla_j \nabla_i X^h + K_{kji}{}^h X^k = 0. \tag{4.1}$$

Transvecting (4.1) with g^{ji}, we find

$$\Box X = 0.$$

Contracting (4.1) with respect to h and i, we obtain

$$\nabla_j(\nabla_i X^i) = 0,$$

from which

$$\delta X = \text{const.}$$

But, since $\int_M (\delta X)\, dV = 0$, we must have

$$\delta X = 0.$$

Thus, by Theorem 3.4, we have

THEOREM 4.1. *An affine Killing vector field in M is a Killing vector field* (Yano [6]).

§5. Projective Killing Vector Fields

A vector field X is a projective Killing vector field if it satisfies

$$(\pounds_X \nabla)(Y, Z) = \pi(Y)Z + \pi(Z)Y \tag{5.1}$$

for any vector fields Y and Z, π being a certain 1-form. From (5.1) we find

$$\Box X = -2P,$$

P being a vector field associated with the 1-form π, and

$$d(\delta X) = -(n+1)\pi.$$

From these two equations, we find

$$\Box X - \frac{2}{n+1} D(\delta X) = 0,$$

or

$$\Delta X - 2QX - \frac{2}{n+1} D(\delta X) = 0,$$

by virtue of (1.8).

Substituting this into (1.12), we find

$$\int_M \left[2R(X, X) - \tfrac{1}{2} |d\xi|^2 - \frac{n-1}{n+1} (\delta X)^2 \right] dV = 0 \tag{5.2}$$

by virtue of

$$\int_M \nabla_X(\delta X)\, dV = \int_M (\delta X)^2\, dV.$$

From (5.2) we see that if the Ricci curvature of M satisfies $R(X, X) \leq 0$ for any X, then

$$d\xi = 0, \qquad \delta X = 0$$

and consequently

$$\Box X = 0 \quad \text{and} \quad \delta X = 0.$$

Hence X is a Killing vector field by Theorem 4.1. Thus, X being harmonic as well as a Killing vector field, its covariant derivative vanishes. Thus we have

THEOREM 5.1. *If the Ricci curvature of M satisfies $R(X, X) \leq 0$ for any X, a projective Killing vector field has vanishing covariant derivative, and if the Ricci curvature is negative-definite, then a projective Killing vector field other than zero does not exist in M* (Couty [2]).

§6. Conformal Killing Vector Fields

A vector field X is a conformal Killing vector field if it satisfies

$$£_X g = 2\rho g, \tag{6.1}$$

where ρ is given by

$$\rho = -\frac{1}{n}\delta X. \tag{6.2}$$

Substituting (6.1) and (6.2) into (1.11) we find

$$\int_M \left[R(X, X) - |\nabla X|^2 - \frac{n-2}{n}(\delta X)^2 \right] dV = 0,$$

from which we have

THEOREM 6.1. *If the Ricci curvature of M satisfies $R(X, X) \leq 0$, a conformal Killing vector field has vanishing covariant derivative, and if the Ricci curvature is negative-definite, a conformal Killing vector field other than the zero vector field does not exist in M* (Yano [6]).

For a conformal Killing vector field X,

$$(£_X\nabla)(Y, Z) = (d\rho)(Y)Z + (d\rho)(Z)Y - g(Y, Z)\, D\rho,$$

from which

$$\square X = (n - 2)\, D\rho,$$

or

$$\square X + \frac{n-2}{n} D\, \delta X = 0 \tag{6.3}$$

by virtue of (6.2).

Conversely, if a vector field X satisfies (6.3), then, from (1.14),

$$\pounds_X g + \frac{2}{n}(\delta X)g = 0,$$

which shows that X is a conformal Killing vector field. Thus we have

THEOREM 6.2. *A necessary and sufficient condition for a vector field X to be a conformal Killing vector field is that*

$$\square X + \frac{n-2}{n} D\, \delta X = 0$$

(Lichnerowicz [4], Satō [1], Yano [6]).

PROBLEMS

1. Prove that if $\delta\, \nabla X = cX$ for a nonzero vector field X, c being a constant, then c is positive.
2. Prove that the integral formula

$$\int_M [K_{ji}X^jX^i + (\nabla^jX^i)(\nabla_iX_j) - (\nabla_jX^j)^2]\, dV = 0$$

can be written as

$$\int_M [K_{ji}X^jX^i - \tfrac{1}{2}(\nabla^jX^i - \nabla^iX^j)(\nabla_jX_i - \nabla_iX_j) + (\nabla^jX^i)(\nabla_jX_i) - (\nabla_jX^j)^2]\, dV = 0$$

or

$$\int_M [K_{ji}X^jX^i + \tfrac{1}{2}(\nabla^jX^i + \nabla^iX^j)(\nabla_jX_i + \nabla_iX_j) - (\nabla^jX^i)(\nabla_jX_i) - (\nabla_jX^j)^2]\, dV = 0,$$

or in index-free notation,

$$\int_M [R(X, X) - \tfrac{1}{2}\, |d\xi|^2 + |\nabla X|^2 - (\delta\xi)^2]\, dV = 0$$

or

$$\int_M [R(X, X) + \tfrac{1}{2}\, |\pounds_X g|^2 - |\nabla X|^2 - (\delta X)^2]\, dV = 0.$$

3. Prove that the integral formula

$$\int_M [g(\square X, X) - \tfrac{1}{2}|\pounds_X g|^2 + (\delta X)^2]\, dV = 0$$

can also be written as

$$\int_M \left[g\left(\square X + \frac{n-2}{n} d\delta X, X \right) - \frac{1}{2} \left| \pounds_X g + \frac{2}{n}(\delta X) g \right|^2 \right] dV = 0.$$

4. Deduce from

$$\pounds_X \Gamma_{ji}^h = \nabla_j \nabla_i X^h + K_{kji}{}^h X^k = \delta_j^h p_i + \delta_i^h p_j$$

the integral formula

$$\int_M \left[2K_{ji} X^j X^i - \tfrac{1}{2}(\nabla^j X^i - \nabla^i X^j)(\nabla_j X_i - \nabla_i X_j) - \frac{n-1}{n+1} (\nabla_j X^j)^2 \right] dV = 0.$$

5. Deduce the equation

$$g^{ji}\, \nabla_j \nabla_i X^h + K_i{}^h X^i - \frac{n-2}{n} \nabla^h (\nabla_i X^i) = 0$$

from

$$\pounds_X g_{ji} = \nabla_j X_i + \nabla_i X_j = 2\rho g_{ji}.$$

CHAPTER

3

Riemannian Manifolds Admitting an Infinitesimal Conformal Transformation

In this chapter we study properties of a compact Riemannian manifold M with constant scalar curvature admitting an infinitesimal conformal transformation. We also assume that M is orientable. If M is not orientable, we have only to take the twofold orientable covering of M.

For Riemannian manifolds admitting a finite conformal transformation, see Barbance [1], Goldberg and Yano [1], Hsiung and Liu [1], Yano and Obata [2].

§1. Preliminaries

We put

$$G_{ji} = K_{ji} - \frac{r}{n} g_{ji}, \tag{1.1}$$

and

$$Z_{kji}{}^{h} = K_{kji}{}^{h} - \frac{r}{n(n-1)} (\delta_k^h g_{ji} - \delta_j^h g_{ki}). \tag{1.2}$$

The tensor G measures the deviation of M from an Einstein manifold and the tensor Z that from a manifold of constant curvature. We can easily verify that

$$G_{ji}g^{ji} = 0 \quad \text{or} \quad \langle G, g\rangle = 0, \tag{1.3}$$

$$Z_{tji}{}^{t} = G_{ji}, \tag{1.4}$$

$$|G|^2 = |R|^2 - \frac{r^2}{n}, \tag{1.5}$$

$$|Z|^2 = |K|^2 - \frac{2r^2}{n(n-1)}, \tag{1.6}$$

and

$$\delta G = -\frac{n-2}{2n}\, dr. \tag{1.7}$$

Thus, for a manifold with constant scalar curvature r, we have

$$\delta G = 0. \tag{1.8}$$

For an infinitesimal conformal transformation X in M, we have

$$£_X g_{ji} = \nabla_j X_i + \nabla_i X_j = 2\rho g_{ji}, \tag{1.9}$$

where

$$\rho = -\frac{1}{n}\,\delta\xi,$$

ξ being the 1-form associated with X. It follows that

$$£_X \left\{ {h \atop ji} \right\} = \delta_j^h \rho_i + \delta_i^h \rho_j - \rho^h g_{ji}, \tag{1.10}$$

where $\rho_i = \nabla_i \rho$. Thus, from the formula

$$£_X K_{kji}{}^h = \nabla_k \left(£_X \left\{ {h \atop ji} \right\} \right) - \nabla_j \left(£_X \left\{ {h \atop ki} \right\} \right),$$

we obtain

$$£_X K_{kji}{}^h = -\delta_k^h \nabla_j \rho_i + \delta_j^h \nabla_k \rho_i - \nabla_k \rho^h g_{ji} + \nabla_j \rho^h g_{ki}, \tag{1.11}$$

from which, by contraction with respect to k and h,

$$£_X K_{ji} = -(n-2)\, \nabla_j \rho_i + (\Delta\rho) g_{ji}. \tag{1.12}$$

Now,

$$£_X r = £_X(K_{ji}g^{ji}) = (£_X K_{ji})g^{ji} + K_{ji}(£_X g^{ji}) = (£_X K_{ji})g^{ji} - 2\rho r$$

by virtue of $£_X g^{ji} = -2\rho g^{ji}$, and consequently, using (1.12),

$$£_X r = 2(n-1)\, \Delta\rho - 2\rho r. \tag{1.13}$$

From (1.9), (1.11), (1.12), and (1.13) we easily obtain

$$£_X G_{ji} = -(n-2)\left(\nabla_j \rho_i + \frac{1}{n}\Delta\rho \cdot g_{ji}\right), \tag{1.14}$$

$$£_X Z_{kji}{}^h = -\delta_k^h \nabla_j \rho_i + \delta_j^h \nabla_k \rho_i - \nabla_k \rho^h g_{ji} + \nabla_j \rho^h g_{ki} - \frac{2}{n}\Delta\rho(\delta_k^h g_{ji} - \delta_j^h g_{ki}). \tag{1.15}$$

§2. Lemmas

In this section we prove some lemmas and integral formulas which will be required later.

LEMMA 2.1. *If a function ρ in M satisfies $\Delta\rho = k\rho$ with a constant k, then*

$$\Delta\, D\rho = kD\rho. \tag{2.1}$$

Proof. This follows from

$$\Delta\, D\rho = D\,\Delta\rho = D\,k\rho = kD\rho.$$

(See Eq. (1.5) in Chapter 2.)

LEMMA 2.2. *If a function ρ in M satisfies $\Delta\rho = k\rho$ with a constant k, then*

$$\int_M \left[R(D\rho, \boldsymbol{D\rho}) - \frac{n-1}{n} k\, |D\rho|^2 + \left| \nabla\nabla\rho + \frac{1}{n} k\rho g \right|^2 \right] dV = 0. \tag{2.2}$$

Proof. The integral formula (1.14) of Chapter 2 can also be written as

$$\int_M \left[g\left(\Delta X - 2QX + \frac{n-2}{n} D\delta X, X \right) - \frac{1}{2}\left| £_X g + \frac{2}{n}(\delta X) g \right|^2 \right] dV = 0.$$

Putting $X = D\rho$ in this formula, we have

$$\int_M \left[g\left(\Delta D\rho - 2QD\rho + \frac{n-2}{n} D\Delta\rho, D\rho \right) \right.$$

$$\left. - \frac{1}{2}\left| 2\nabla\nabla\rho + \frac{2}{n}(\Delta\rho) g \right|^2 \right] dV = 0,$$

and using (2.1),

$$\int_M \left[g\left(kD\rho - 2QD\rho + \frac{n-2}{n} kD\rho, D\rho \right) - 2\left| \nabla\nabla\rho + \frac{1}{n} k\rho g \right|^2 \right] dV = 0,$$

or

$$\int_M \left[R(D\rho, D\rho) - \frac{n-1}{n} k\, |D\rho|^2 + \left| \nabla\nabla\rho + \frac{1}{n} k\rho g \right|^2 \right] dV = 0.$$

LEMMA 2.3. *If M has constant scalar curvature r and admits an infinitesimal nonisometric conformal transformation X: $\pounds_X g = 2\rho g$, $\rho \neq 0$, then*

$$\Delta\rho = \frac{r}{n-1} \rho \qquad (r > 0), \tag{2.3}$$

and consequently

$$\int_M \rho \, dV = 0. \tag{2.4}$$

Proof. Equation (2.3) follows from (1.13), in which $r = \text{const}$, and Eq. (2.4) follows from (2.3) by integration.

LEMMA 2.4. *Under the same assumptions as in Lemma 2.3,*

$$\int_M |d\rho|^2 \, dV = \frac{r}{n-1} \int \rho^2 \, dV. \tag{2.5}$$

Proof. This follows from

$$\begin{aligned} \tfrac{1}{2}\Delta\rho^2 &= (\Delta\rho)\rho - |d\rho|^2 \\ &= \frac{r}{n-1} \rho^2 - |d\rho|^2 \end{aligned}$$

by integration.

LEMMA 2.5. *Under the same assumptions as in Lemma 2.3, the scalar curvature r is positive.*

Proof. This follows from Lemma 2.4.

§3. Integral Formulas

We prove in this section two useful integral formulas.

LEMMA 3.1. *If a compact and orientable Riemannian manifold M of dimension greater than 2 has constant scalar curvature r and admits an infinitesimal nonisometric conformal transformation X: $£_X g = 2\rho g$, $\rho \neq 0$, then*

$$\int_M G(D\rho, D\rho)\, dV = \frac{1}{n-2}\int_M [2\rho^2 |G|^2 + \tfrac{1}{2}\rho £_X |G|^2]\, dV. \tag{3.1}$$

Proof. To prove this integral formula, we first compute $£_X |G|^2$. We have

$$\begin{aligned} £_X |G|^2 &= 2\langle £_X G, G\rangle - 4\rho\, |G|^2 \\ &= -2(n-2)\left\langle \nabla\nabla\rho + \frac{1}{n}\Delta\rho \cdot g, G\right\rangle - 4\rho\, |G|^2 \end{aligned}$$

by virtue of (1.14), from which

$$\langle G, \nabla\nabla\rho\rangle = -\frac{2}{n-2}\rho\, |G|^2 - \frac{1}{2(n-2)} £_X |G|^2 \tag{3.2}$$

because $\langle G, g\rangle = 0$.

On the other hand,

$$-\delta\{G(\rho D\rho)\} = G(D\rho, D\rho) + \rho\langle G, \nabla\nabla\rho\rangle \tag{3.3}$$

by virtue of $\delta G = 0$, where, in $G(\rho D\rho)$, we regard G as a linear transformation field acting on the vector field $\rho D\rho$ and giving a 1-form $G(\rho D\rho)$. Thus, using (3.2),

$$-\delta\{G(\rho D\rho)\} = G(D\rho, D\rho) - \frac{2}{n-2}\rho^2 |G|^2 - \frac{1}{2(n-2)}\rho £_X |G|^2.$$

Integrating this equation over the manifold M, we obtain (3.1).

LEMMA 3.2. *Under the same assumptions as in Lemma 3.1,*

$$\int_M G(D\rho, D\rho)\, dV = \int_M [\tfrac{1}{2}\rho^2 |Z|^2 + \tfrac{1}{8}\rho £_X |Z|^2]\, dV. \tag{3.4}$$

Proof. To prove this integral formula, we first compute $£_X |Z|^2$. We have

$$\begin{aligned} £_X |Z|^2 &= 2\langle £_X Z, Z\rangle - 4\rho\, |Z|^2 \\ &= -8\langle \nabla\nabla\rho, G\rangle - 4\rho\, |Z|^2 \end{aligned}$$

by virtue of (1.15), from which

$$\langle G, \nabla\nabla\rho\rangle = -\tfrac{1}{2}\rho\, |Z|^2 - \tfrac{1}{8}£_X |Z|^2.$$

Thus, from (3.3),

$$-\delta(G\rho D\rho) = G(D\rho, D\rho) - \tfrac{1}{2}\rho^2 |Z|^2 - \tfrac{1}{8}\rho £_X |Z|^2.$$

Integrating this equation over the manifold M, we obtain (3.4).

§4. Theorems on Riemannian Manifolds with Constant Scalar Curvature and Admitting an Infinitesimal Nonisometric Conformal Transformation

In this section we need the following theorem, which is due to Obata [1, 3, 4]:

THEOREM 4.1. *If a complete Riemannian manifold M of dimension $n \geq 2$ admits a nonconstant function ρ such that*

$$\nabla\nabla\rho = -c^2\rho g,$$

where c is a positive constant, then M is isometric to a sphere of radius $1/c$ in $(n + 1)$-dimensional Euclidean space.

We first prove

THEOREM 4.2. *If a compact Riemannian manifold M of dimension $n \geq 2$ with $r =$ const admits an infinitesimal nonisometric conformal transformation X: $£_X g = 2\rho g$, $\rho \neq 0$, and if one of the following conditions is satisfied, then M is isometric to a sphere.*

(*1*) *The 1-form ξ associated with X is the differential of a function.*

(*2*) *$QD\rho = \lambda D\rho$, where λ is a constant.*

(*3*) *$£_X R = \alpha g$, where α is a function.*

(Bishop and Goldberg [1]. Lichnerowicz [11]. Yano and Obata [1].)

Proof. From Lemma 2.3,

$$\Delta\rho = \frac{r}{n-1}\rho. \tag{4.1}$$

On the other hand, ξ being the differential of a function f, we have, from $£_X g = 2\rho g$,

$$\nabla\nabla f = \rho g, \tag{4.2}$$

from which,

$$\Delta f = -n\rho. \tag{4.3}$$

Thus, from (4.1) and (4.3),

$$\Delta\rho = -\frac{r}{n(n-1)}\Delta f,$$

or

$$\Delta\left(\rho + \frac{r}{n(n-1)} f\right) = 0,$$

from which, by Proposition 1.2 of Chapter 2,

$$\rho + \frac{r}{n(n-1)} f = \text{const.}$$

Thus,

$$\nabla\nabla\rho + \frac{r}{n(n-1)} \nabla\nabla f = 0,$$

or, by (4.2),

$$\nabla\nabla\rho = -\frac{r}{n(n-1)} \rho g.$$

Since ρ is not a constant function, M is isometric to a sphere by the theorem of Obata. This proves the first proposition.

We next assume that

$$QD\rho = \lambda D\rho.$$

Regarding Q as a linear endomorphism of the space of 1-forms, we have

$$\begin{aligned} \delta\{Q(\rho\xi)\} &= -\langle R, d\rho \otimes \xi\rangle - \rho\langle R, \nabla\xi\rangle \\ &= -\lambda(d\rho)(X) - \tfrac{1}{2}\rho\langle R, \pounds_X g\rangle \\ &= \lambda\delta(\rho\xi) - \lambda\rho\delta\xi - r\rho^2. \end{aligned}$$

Consequently,

$$\delta\{Q(\rho\xi)\} = \lambda\delta(\rho\xi) + (n\lambda - r)\rho^2$$

by virtue of $\delta Q = 0$, $\pounds_X g = 2\rho g$ and $\delta\xi = -n\rho$, from which, integrating over M,

$$\int_M (n\lambda - r)\rho^2 \, dV = 0.$$

Hence

$$\lambda = \frac{r}{n}.$$

Substituting

$$k = \frac{r}{n-1}, \qquad R(D\rho, D\rho) = \frac{r}{n} |D\rho|^2$$

into (2.2), we find

$$\int \left| \nabla\nabla\rho + \frac{r}{n(n-1)} \rho g \right|^2 dV = 0,$$

from which

$$\nabla\nabla\rho = -\frac{1}{n(n-1)} r\rho g,$$

which proves the second proposition.

Assume now that

$$£_X R = \alpha g,$$

α being a function. Then, from (1.12),

$$-(n-2)\,\nabla\nabla\rho + (\Delta\rho)g = \alpha g,$$

from which

$$\alpha = \frac{2(n-1)}{n}\Delta\rho.$$

Consequently,

$$-(n-2)\left(\nabla\nabla\rho + \frac{1}{n}\Delta\rho g\right) = 0,$$

that is,

$$\nabla\nabla\rho = -\frac{1}{n(n-1)} r\rho g,$$

which proves the third proposition.

We next prove

THEOREM 4.3. *If a compact Riemannian manifold M of dimension n > 2 with scalar curvature* $r =$ const *admits an infinitesimal nonisometric conformal transformation X*: $£_X g = 2\rho g$, $\rho \neq$ const, *such that*

$$\int_M G(D\rho, D\rho)\,dV \geq 0, \tag{4.4}$$

then M is isometric to a sphere (Yano [17]).

Proof. Since $r =$ const we have, from (1.13),

$$\Delta\rho = \frac{1}{n-1} r\rho \quad (r > 0).$$

Thus, from Lemma 2.2,

$$\int_M \left[G(D\rho, D\rho) + \left|\nabla\nabla\rho + \frac{1}{n(n-1)} r\rho g\right|^2\right] dV = 0, \tag{4.5}$$

k being equal to $(1/(n-1))r$, from which, $\int_M G(D\rho, D\rho)\,dV$ being nonnegative,

$$\nabla\nabla\rho = -\frac{1}{n(n-1)} r\rho g.$$

Consequently, by Theorem 4.1, M is isometric to a sphere.

COROLLARY 4.4. *In order that a compact Riemannian manifold M of dimension $n > 2$ with scalar curvature $r =$ const and admitting an infinitesimal nonisometric conformal transformation X: $\pounds_X g = 2\rho g$, $\rho \neq$ const, be isometric to a sphere, it is necessary and sufficient that*

$$\int_M G(D\rho, D\rho)\, dV = 0 \tag{4.6}$$

(Goldberg [4], Weber and Goldberg [1], Yano [19]).

Proof. Suppose that M is isometric to a sphere; then $G = 0$ and consequently we have (4.6).

Conversely, if (4.6) holds, then by Theorem 4.3 M is isometric to a sphere.

COROLLARY 4.5. *If a compact Einstein manifold M of dimension $n > 2$ admits an infinitesimal nonisometric conformal transformation, then M is isometric to a sphere* (Yano and Nagano [1]).

Proof. For an Einstein manifold M, we have

$$G = 0$$

and (4.6) is always satisfied.

We next prove

THEOREM 4.6. *If a compact Riemannian manifold M of dimension $n > 2$ with scalar curvature $r =$ const admits an infinitesimal nonisometric conformal transformation X such that*

$$\pounds_X |R|^2 = 0, \tag{4.7}$$

then M is isometric to a sphere.

Proof. We first note that, by (1.5), $\pounds_X |R|^2 = 0$ is equivalent to

$$\pounds_X |G|^2 = 0.$$

Thus, if $\pounds_X |G|^2 = 0$, then we have, from (3.1),

$$\int_M G(D\rho, D\rho)\, dV = \frac{2}{n-2}\int_M \rho^2 |G|^2\, dV \geq 0,$$

and consequently, by Theorem 4.3, M is isometric to a sphere.

The following corollaries follow immediately from Theorem 4.6.

COROLLARY 4.7. *If a compact Riemannian manifold of dimension $n > 2$, with $r =$ const and $|R|^2 =$ const, admits an infinitesimal nonisometric conformal transformation, then M is isometric to a sphere* (Lichnerowicz [11]).

COROLLARY 4.8. *If a compact homogeneous Riemannian manifold M of dimension $n > 2$ admits an infinitesimal nonisometric conformal transformation, then M is isometric to a sphere* (Goldberg and Kobayashi [1–3]).

We now prove

THEOREM 4.9. *If a compact Riemannian manifold M of dimension $n > 2$ with scalar curvature $r =$ const admits an infinitesimal nonisometric conformal transformation X such that*

$$\pounds_X |K|^2 = 0, \tag{4.8}$$

then M is isometric to a sphere.

Proof. We first note that, by (1.6), $\pounds_X |K|^2 = 0$ is equivalent to

$$\pounds_X |Z|^2 = 0.$$

Thus, if $\pounds_X |Z|^2 = 0$, then from (3.4)

$$\int_M G(D\rho, D\rho)\, dV = \frac{1}{2}\int_M \rho^2 |Z|^2\, dV \geq 0,$$

and consequently, by Theorem 4.3, M is isometric to a sphere.

From Theorem 4.9, we have

COROLLARY 4.10. *If a compact Riemannian manifold M of dimension $n > 2$, with $r =$ const and $|K|^2 =$ const, admits an infinitesimal nonisometric conformal transformation, then M is isometric to a sphere* (Hsiung [7]).

THEOREM 4.11. *If M is complete, of dimension $n > 2$, with $r = const > 0$, and if it admits an infinitesimal nonisometric conformal transformation X: $\pounds_X g = 2\rho$, $\rho \neq$ const, then*

$$\rho^2 r^2 \leq n(n-1)^2 |\nabla\nabla\rho|^2, \tag{4.9}$$

equality holding if and only if M is isometric to a sphere (Goldberg [4], Yano [19]).

Proof. We know that the function ρ satisfies

$$\Delta\rho = \frac{1}{n-1} r\rho,$$

and consequently, we have

$$\begin{aligned} 0 &\leq \left| \nabla\nabla\rho + \frac{1}{n(n-1)} r\rho g \right|^2 \\ &= |\nabla\nabla\rho|^2 - \frac{2}{n(n-1)} r\rho\,\Delta\rho + \frac{1}{n(n-1)^2} r^2\rho^2 \\ &= |\nabla\nabla\rho|^2 - \frac{1}{n(n-1)^2} r^2\rho^2, \end{aligned}$$

from which

$$\rho^2 r^2 \leq n(n-1)^2 \, |\nabla\nabla\rho|^2,$$

equality holding if and only if

$$\nabla\nabla\rho = -\frac{1}{n(n-1)} r\rho g,$$

that is, if and only if M is isometric to a sphere.

THEOREM 4.12. *If M is compact, of dimension $n > 2$, with $r =$ const, and if it admits an infinitesimal nonisometric conformal transformation X: $\pounds_X g = 2\rho g$, $\rho \neq$ const, then*

$$n(n-1) \frac{\int_M R(D\rho, D\rho)\, dV}{\int_M \rho^2\, dV} \leq r^2, \tag{4.10}$$

equality holding if and only if M is isometric to a sphere (Yano [19]).

Proof. From (4.5) we have

$$\int_M G(D\rho, D\rho)\, dV \leq 0, \tag{4.11}$$

equality holding if and only if

$$\nabla\nabla\rho = -\frac{1}{n(n-1)} r\rho g,$$

that is, if and only if M is isometric to a sphere. But (4.11) can be written as

$$\int_M R(D\rho, D\rho)\, dV - \frac{1}{n} \int_M r\, |\nabla\rho|^2\, dV \leq 0,$$

from which

$$\int_M R(D\rho, D\rho)\, dV - \frac{r^2}{n(n-1)} \int_M \rho^2\, dV \leq 0,$$

or

$$n(n-1) \frac{\int_M R(D\rho, D\rho)\, dV}{\int_M \rho^2\, dV} \leq r^2,$$

since

$$\int_M |\nabla \rho|^2\, dV = \frac{r}{n-1} \int_M \rho^2\, dV$$

by (2.5).

PROBLEMS

1. Put

$$G_{ji} = K_{ji} - \frac{r}{n} g_{ji},$$

$$Z_{kji}{}^h = K_{kji}{}^h - \frac{r}{n(n-1)} (\delta_k^h g_{ji} - \delta_j^h g_{ki}).$$

Prove that

$$G_{ji} g^{ji} = 0,$$

$$Z_{tji}{}^t = G_{ji},$$

$$|G|^2 = |R|^2 - \frac{r^2}{n},$$

$$|Z|^2 = |K|^2 - \frac{2r^2}{n(n-1)},$$

and

$$\delta G = -\frac{n-2}{2n} dr.$$

2. Prove that, for a conformal Killing vector field X: $\pounds_X g = 2\rho g$,

$$\pounds_X G_{ji} = -(n-2)\left(\nabla_j \rho_i + \frac{1}{n} \Delta\rho \cdot g_{ji} \right),$$

$$\pounds_X Z_{kji}{}^h = -\delta_k^h \nabla_j \rho_i + \delta_j^h \nabla_k \rho_i - \nabla_k \rho^h g_{ji} + \nabla_j \rho^h g_{ki} - \frac{2}{n} \Delta\rho (\delta_k^h g_{ji} - \delta_j^h g_{ki}).$$

3. Prove that, for a conformal Killing vector field X: $\pounds_X g = 2\rho g$,

$$\pounds_X |G|^2 = -2(n-2) \left\langle \nabla\nabla\rho + \frac{1}{n} \Delta\rho \cdot g, G \right\rangle - 4\rho |G|^2,$$

and consequently

$$\langle G, \nabla\nabla\rho\rangle = -\frac{2}{n-2}\,\rho\,|G|^2 - \frac{1}{2(n-2)}\,£_X\,|G|^2.$$

4. Prove that, for a conformal Killing vector field X: $£_X g = 2\rho g$,

$$£_X\,|Z|^2 = 8\left\langle \nabla\nabla\rho + \frac{1}{n}\,\Delta\rho\cdot g, G\right\rangle - 4\rho\,|Z|^2,$$

and consequently

$$\langle G, \nabla\nabla\rho\rangle = -\tfrac{1}{2}\rho\,|Z|^2 - \tfrac{1}{8}£_X\,|Z|^2.$$

5. Consider an $(n+1)$-dimensional Euclidean space E^n and denote by X the so-called position vector. Let

$$X = X(x^1, x^2, \ldots, x^n)$$

be a parametric representation of the n-sphere S^n with center at the origin and with radius r. Then,

$$X\cdot X = r^2$$

identically where the "·" represents the inner product. Consequently,

$$X_i\cdot X = 0,$$

where

$$X_i = \partial_i X$$

are n linearly independent vectors tangent to S^n. The fundamental metric tensor g_{ji} of S^n is given by

$$g_{ji} = X_j\cdot X_i.$$

Its unit normal N is given by

$$N = \frac{1}{r}\,X. \qquad (*)$$

The equations of Gauss and Weingarten for S^n can be written as

$$\begin{cases} \partial_j X_i = \left\{ {h \atop ji} \right\} X_h + h_{ji}N, \\ \partial_j N = -h_j{}^i X_i, \end{cases}$$

where the $\left\{ {h \atop ji} \right\}$ are the Christoffel symbols formed with the g_{ji}, h_{ji}, the components of the second fundamental tensor, and $h_j{}^i = h_{jt}g^{ti}$.

From (*)

$$\partial_j N = \frac{1}{r}\,X_j,$$

so that

$$h_j{}^i = -\frac{1}{r}\,\delta_j^i,$$

and consequently

$$h_{ji} = -\frac{1}{r}\,g_{ji}.$$

Thus the equations of Gauss and Weingarten for the sphere can be written as

$$\begin{cases} \nabla_j X_i = \partial_j X_i - \begin{Bmatrix} h \\ ji \end{Bmatrix} X_h = -\dfrac{1}{r} g_{ji} N, \\ \nabla_j N = \partial_j N = \dfrac{1}{r} X_j. \end{cases}$$

Prove that if we take an arbitrary but fixed unit vector A and put

$$f = A \cdot X,$$

then the function f satisfies

$$\nabla_j \nabla_i f = -\frac{1}{r^2} f g_{ji}.$$

6. Use Corollary 4.5 to prove that if a compact Riemannian manifold M of constant curvature of dimension $n > 2$ admits an infinitesimal nonisometric conformal transformation, then M is isometric to a sphere (Bishop and Goldberg [1]).
7. Prove that if a compact Riemannian manifold M of dimension $n > 2$ with scalar curvature r = const admits an infinitesimal nonisometric conformal transformation X: $\pounds_X g = 2\rho g$, $\rho \neq$ const, such that one of the following conditions is satisfied, then M is isometric to a sphere:
   (1) $4\rho |G|^2 + \pounds_X |G|^2 = 0$,
   (2) $\pounds_X |G|^2 = k\rho |G|^2$,
   where $k \geq -4$ is a constant,
   (3) $\pounds_X |G|^2 = k\rho^{2t+1} |G|^2$,
   where $k > 0$ and t is an integer (Yano and Obata [2]).
8. Prove that in Problem 7 the conditions (1–3) can be replaced by
   (1′) $4\rho |Z|^2 + \pounds_X |Z|^2 = 0$,
   (2′) $\pounds_X |Z|^2 = k\rho |Z|^2$,
   respectively, where $k \geq -4$ is a constant,
   (3′) $\pounds_X |Z|^2 = k\rho^{2t+1} |Z|^2$,
   where $k > 0$ and t is an integer.

CHAPTER

4

Harmonic Forms and Killing Tensor Fields

In Chapter 2 we studied harmonic 1-forms and Killing vector fields. In this chapter we study harmonic p-forms ($p \geq 1$) and Killing tensor fields in compact and orientable Riemannian manifold. We will assume in the sequel that M is compact and orientable.

§1. Harmonic p-Forms

Consider a p-form

$$\omega = \frac{1}{p!} \omega_{i_1 i_2 \cdots i_p} \, dx^{i_1} \wedge dx^{i_2} \wedge \cdots \wedge dx^{i_p}, \tag{1.1}$$

the coefficients being components of a skew-symmetric tensor field of type $(0, p)$.

The differential $d\omega$ of ω is a $(p+1)$-form given by

$$d\omega = \frac{1}{(p+1)!}(\partial_i \omega_{i_1 i_2 \cdots i_p} - \partial_{i_1} \omega_{i i_2 \cdots i_p} - \cdots - \partial_{i_p} \omega_{i_1 \cdots i_{p-1} i}) \, dx^i \wedge dx^{i_1} \wedge \cdots \wedge dx^{i_p}, \tag{1.2}$$

or

$$d\omega = \frac{1}{(p+1)!}(\nabla_i \omega_{i_1 i_2 \cdots i_p} - \nabla_{i_1} \omega_{i i_2 \cdots i_p} - \cdots - \nabla_{i_p} \omega_{i_1 \cdots i_{p-1} i}) \, dx^i \wedge dx^{i_1} \wedge \cdots \wedge dx^{i_p}, \tag{1.3}$$

the coefficients being components of a skew-symmetric tensor field of type $(0, p + 1)$.

The codifferential $\delta\omega$ of ω is a $(p - 1)$-form given by

$$\delta\omega = -\frac{1}{(p-1)!} g^{ji}\nabla_j \omega_{ii_2\cdots i_p}\, dx^{i_2} \wedge \cdots \wedge dx^{i_p}, \tag{1.4}$$

the coefficients being components of a skew-symmetric tensor field of type $(0, p - 1)$. The codifferential of a function is defined to be zero.

It is easily checked that

$$d(d\omega) = 0, \qquad \delta(\delta\omega) = 0. \tag{1.5}$$

A *harmonic form* is an exterior differential form ω which satisfies

$$d\omega = 0 \quad \text{and} \quad \delta\omega = 0,$$

and a harmonic tensor field is a tensor field whose components are the coefficients of a harmonic form.

We now consider a p-form

$$u = \frac{1}{p!} u_{i_1 i_2\cdots i_p}\, dx^{i_1} \wedge dx^{i_2} \wedge \cdots \wedge dx^{i_p}$$

and a $(p + 1)$-form

$$v = \frac{1}{(p+1)!} v_{ii_1 i_2\cdots i_p}\, dx^{i} \wedge dx^{i_1} \wedge \cdots \wedge dx^{i_p},$$

and construct the vector field whose components are

$$u_{i_1 i_2\cdots i_p} v^{ii_1\cdots i_p},$$

the $v^{ii_1\cdots i_p}$ being contravariant components of the tensor field defined by v.

Applying Green's theorem to this vector field, we find

$$\int_M \nabla_j (u_{i_1 i_2\cdots i_p} v^{ji_1\cdots i_p})\, dV = 0,$$

or

$$\int_M \frac{1}{(p+1)!} (\nabla_j u_{i_1 i_2\cdots i_p} - \nabla_{i_1} u_{ji_2\cdots i_p} - \cdots - \nabla_{i_p} u_{i_1\cdots i_p j}) v^{ji_1 i_2\cdots i_p}\, dV$$
$$+ \int_M \frac{1}{p!} u_{i_1 i_2\cdots i_p} \nabla_j v^{ji_1\cdots i_p}\, dV = 0.$$

Thus, if we introduce the notation

$$\int_M \frac{1}{p!} a_{i_1 i_2\cdots i_p} b^{i_1 i_2\cdots i_p}\, dV = (a, b) \tag{1.6}$$

for the integral of the scalar product of two skew-symmetric tensor fields $a_{i_1\cdots i_p}$ and $b_{i_1\cdots i_p}$ of the same type, $(0, p)$, or two exterior differential forms

$$a = \frac{1}{p!} a_{i_1 i_2 \cdots i_p} dx^{i_1} \wedge dx^{i_2} \wedge \cdots \wedge dx^{i_p}$$

and

$$b = \frac{1}{p!} b_{i_1 i_2 \cdots i_p} dx^{i_1} \wedge dx^{i_2} \wedge \cdots \wedge dx^{i_p},$$

then,

$$(du, v) = (u, \delta v).$$

Hence,

PROPOSITION 1.1. *For a p-form u and a $(p + 1)$-form v in a compact orientable Riemannian manifold,*

$$(du, v) = (u, \delta v). \tag{1.7}$$

Take a p-form ω and put

$$u = \omega, \qquad v = d\omega,$$

then, by Proposition 1.1,

$$(\delta\, d\omega, \omega) = (d\omega, d\omega).$$

Next, put

$$u = \delta\omega, \qquad v = \omega;$$

then

$$(d\, \delta\omega, \omega) = (\delta\omega, \delta\omega).$$

Thus,

$$(\Delta\omega, \omega) = (d\omega, d\omega) + (\delta\omega, \delta\omega),$$

where

$$\Delta\omega = (\delta d + d\delta)\omega. \tag{1.8}$$

PROPOSITION 1.2. *For a p-form ω in a compact and orientable Riemannian manifold,*

$$(\delta\, d\omega, \omega) = (d\omega, d\omega), \tag{1.9}$$

$$(d\, \delta\omega, \omega) = (\delta\omega, \delta\omega), \tag{1.10}$$

and

$$(\Delta\omega, \omega) = (d\omega, d\omega) + (\delta\omega, \delta\omega). \tag{1.11}$$

Suppose that a harmonic p-form ω is *derived*, that is, ω is the exterior differential of a $(p - 1)$-form u, that is,

$$\omega = du,$$

then, ω being harmonic,

$$\delta\omega = \delta\, du = 0.$$

Thus, putting $\omega = u$ in (1.9), we get

$$(du, du) = 0,$$

from which

$$\omega = du = 0.$$

PROPOSITION 1.3. *If a harmonic form is derived, it is identically zero.*

If a p-form ω is the codifferential of a $(p+1)$-form v, that is,

$$\omega = \delta v,$$

then ω is said to be *coderived.*

We can then prove similarly

PROPOSITION 1.4. *If a harmonic form is coderived, it is identically zero.*

Suppose that ω is a harmonic p-form. Then

$$d\omega = 0, \qquad \delta\omega = 0,$$

and consequently

$$\Delta\omega = (\delta d + d\delta)\omega = 0.$$

Conversely, if we have $\Delta\omega = 0$, then, from (1.11), we have

$$(d\omega, d\omega) + (\delta\omega, \delta\omega) = 0,$$

from which

$$d\omega = 0, \qquad \delta\omega = 0,$$

and consequently ω is a harmonic p-form.

THEOREM 1.5. *A necessary and sufficient condition for a p-form ω to be harmonic is*

$$\Delta\omega = 0.$$

We now compute the components of $\Delta\omega$. Let

$$\omega = \frac{1}{p!}\,\omega_{i_1 i_2\cdots i_p}\, dx^{i_1} \wedge dx^{i_2} \wedge \cdots \wedge dx^{i_p}$$

be a local expression for ω. Then

$$d\omega = \frac{1}{(p+1)!}(\nabla_i \omega_{i_1 i_2\cdots i_p} - \nabla_{i_1}\omega_{i i_2\cdots i_p} - \cdots \\ - \nabla_{i_p}\omega_{i_1 i_2\cdots i_{p-1} i})\, dx^i \wedge dx^{i_1} \wedge \cdots \wedge dx^{i_p},$$

and consequently

$$\delta d\omega = -\frac{1}{p!} g^{ji}\nabla_j(\nabla_i\omega_{i_1i_2\cdots i_p} - \nabla_{i_1}\omega_{ii_2\cdots i_p} - \cdots - \nabla_{i_p}\omega_{i_1i_2\cdots i_{p-1}i})\, dx^{i_1}\wedge dx^{i_2}\wedge\cdots\wedge dx^{i_p}.$$

On the other hand,

$$\delta\omega = -\frac{1}{(p-1)!} g^{ji}\nabla_j\omega_{ii_2\cdots i_p}\, dx^{i_2}\wedge\cdots\wedge dx^{i_p},$$

and consequently

$$d\delta\omega = -\frac{1}{p!}(g^{ji}\nabla_{i_1}\nabla_j\omega_{ii_2\cdots i_p} - g^{ji}\nabla_{i_2}\nabla_j\omega_{ii_1i_3\cdots i_p} - \cdots - g^{ji}\nabla_{i_p}\nabla_j\omega_{ii_2\cdots i_{p-1}i_1})\, dx^{i_1}\wedge\cdots\wedge dx^{i_p}$$

Thus,

$$\begin{aligned}
\Delta\omega &= \delta\, d\omega + d\,\delta\omega \\
&= -\frac{1}{p!}\Big\{g^{ji}\nabla_j\nabla_i\omega_{i_1\cdots i_p} \\
&\quad - g^{ji}(\nabla_j\nabla_{i_1}\omega_{ii_2\cdots i_p} - \nabla_{i_1}\nabla_j\omega_{ii_2\cdots i_p}) \\
&\quad - g^{ji}(\nabla_j\nabla_{i_2}\omega_{i_1i\cdots i_p} - \nabla_{i_2}\nabla_j\omega_{i_1i\cdots i_p}) \\
&\quad \cdots \\
&\quad - g^{ji}(\nabla_j\nabla_{i_p}\omega_{i_1\cdots i_{p-1}i} - \nabla_{i_p}\nabla_j\omega_{i_1\cdots i_{p-1}i})\Big\}\, dx^{i_1}\wedge\cdots\wedge dx^{i_p} \\
&= -\frac{1}{p!}\Big\{g^{ji}\nabla_j\nabla_i\omega_{i_1\cdots i_p} \\
&\quad + g^{ji}(K_{ji_1i}{}^t\omega_{ti_2\cdots i_p} + K_{ji_1i_2}{}^t\omega_{it\cdots i_p} + \cdots + K_{ji_1i_p}{}^t\omega_{ii_2\cdots t}) \\
&\quad + g^{ji}(K_{ji_2i_1}{}^t\omega_{ti\cdots i_p} + K_{ji_2i}{}^t\omega_{i_1t\cdots i_p} + \cdots + K_{ji_2i_p}{}^t\omega_{i_1i\cdots t}) \\
&\quad \cdots \\
&\quad + g^{ji}(K_{ji_pi_1}{}^t\omega_{ti_2\cdots i} + K_{ji_pi_2}{}^t\omega_{i_1t\cdots i_p} \\
&\quad + \cdots + K_{ji_pi}{}^t\omega_{i_1\cdots t})\Big\}\, dx^{i_1}\wedge\cdots\wedge dx^{i_p} \\
&= -\frac{1}{p!}\Big\{g^{ji}\nabla_j\nabla_i\omega_{i_1\cdots i_p} - \sum_{s=1}^{p} K_{i_s}{}^t\omega_{i_1\cdots t\cdots i_p} \\
&\quad - \sum_{t<s}^{1\cdots p} K_{i_ti_s}{}^{vu}\omega_{i_1\cdots v\cdots u\cdots i_p}\Big\}\, dx^{i_1}\wedge\cdots\wedge dx^{i_p},
\end{aligned}$$

by virtue of the Ricci identity and

$$g^{ji}K_{ji_si}{}^{t} = -K_{i_s}{}^{t}.$$

Thus we have

THEOREM 1.6. *A necessary and sufficient condition for a p-form*

$$\omega = \frac{1}{p!}\omega_{i_1i_2\cdots i_p}\,dx^{i_1}\wedge dx^{i_2}\wedge\cdots\wedge dx^{i_p}$$

in M to be harmonic is

$$g^{ji}\nabla_j\nabla_i\omega_{i_1\cdots i_p} - \sum_{s=1}^{p} K_{i_s}{}^{t}\omega_{i_1\cdots t\cdots i_p} - \sum_{t<s}^{1\cdots p} K_{i_ti_s}{}^{vu}\omega_{i_1\cdots v\cdots u\cdots i_p} = 0. \tag{1.12}$$

§2. Killing Tensor Fields

We have defined a Killing form as an exterior differential form

$$\omega = \frac{1}{p!}\omega_{i_1i_2\cdots i_p}\,dx^{i_1}\wedge\cdots\wedge dx^{i_p}$$

which satisfies

$$(p+1)\,\nabla\omega = d\omega. \tag{2.1}$$

This means that $\nabla\omega$ is skew-symmetric and that

$$\delta\omega - 0. \tag{2.2}$$

A Killing tensor field is one whose components are coefficients of a Killing form.

Suppose that a Killing p-form ω is derived, that is, $\omega = du$, then, ω being a Killing form, $\delta\, du = 0$. Thus, putting $\omega = u$ in (1.9), we have

$$(du, du) = 0$$

and consequently

$$\omega = du = 0.$$

Hence,

PROPOSITION 2.1. *If a Killing form is derived, it is identically zero.*

Suppose that ω is a Killing p-form. Then

$$d\omega = (p+1)\,\nabla\omega, \qquad \delta\omega = 0,$$

and consequently

$$\begin{aligned}\Delta\omega &= \delta\, d\omega + d\,\delta\omega\\ &= (p+1)\,\delta\,\nabla\omega.\end{aligned}$$

Conversely, if we have

$$\Delta\omega = (p + 1)\,\delta\,\nabla\omega, \qquad \delta\omega = 0,$$

then, from (1.11),

$$(p + 1)(\delta\,\nabla\omega, \omega) = (d\omega, d\omega),$$

from which

$$(p + 1)(\nabla\omega, d\omega) = (d\omega, d\omega)$$

or

$$(p + 1)^2(\nabla\omega, \nabla\omega) = (d\omega, d\omega). \tag{2.3}$$

Thus,

$$\begin{aligned}((p + 1)\,\nabla\omega - d\omega, (p + 1)\,\nabla\omega - d\omega)&\\ &= (p + 1)^2(\nabla\omega, \nabla\omega) - 2(p + 1)(\nabla\omega, d\omega) + (d\omega, d\omega)\\ &= (p + 1)^2(\nabla\omega, \nabla\omega) - 2(d\omega, d\omega) + (d\omega, d\omega)\\ &= (p + 1)^2(\nabla\omega, \nabla\omega) - (d\omega, d\omega)\\ &= 0,\end{aligned}$$

from which

$$(p + 1)\,\nabla\omega = d\omega,$$

and consequently ω is a Killing p-form. Hence,

THEOREM 2.2. *A necessary and sufficient condition for a p-form ω in M to be a Killing p-form is*

$$\Delta\omega = (p + 1)\,\delta\,\nabla\omega, \qquad \delta\omega = 0. \tag{2.4}$$

We now compute the components of $\Delta\omega - (p + 1)\,\delta\,\nabla\omega$. We know that the components of $\Delta\omega$ are

$$-\frac{1}{p!}\left\{g^{ji}\nabla_j\nabla_i\omega_{i_1i_2\cdots i_p} - \sum_{s=1}^{p} K_{i_s}{}^t\omega_{i_1\cdots t\cdots i_p} - \sum_{t<s}^{1\cdots p} K_{i_ti_s}{}^{vu}\omega_{i_1\cdots v\cdots u\cdots i_p}\right\}$$

and those of $(p + 1)\,\delta\,\nabla\omega$ are

$$-\frac{p+1}{p!}\,g^{ji}\nabla_j\nabla_i\omega_{i_1i_2\cdots i_p},$$

and consequently the components of $\Delta\omega - (p + 1)\,\delta\,\nabla\omega$ are given by

$$\frac{1}{(p-1)!}\left\{g^{ji}\nabla_j\nabla_i\omega_{i_1i_2\cdots i_p} + \frac{1}{p}\sum_{s=1}^{p} K_{i_s}{}^t\omega_{i_1\cdots t\cdots i_p} + \frac{1}{p}\sum_{t<s}^{1\cdots p} K_{i_ti_s}{}^{vu}\omega_{i_1\cdots v\cdots u\cdots i_p}\right\}.$$

On the other hand, the components of $\delta\omega$ are given by

$$-\frac{1}{(p-1)!}\{g^{ji}\nabla_j\omega_{ii_2\cdots i_p}\}.$$

Thus Theorem 2.2, in terms of coordinates, is given by

THEOREM 2.3. *A necessary and sufficient condition for a p-form*

$$\omega = \frac{1}{p!}\omega_{i_1 i_2\cdots i_p}\,dx^{i_1}\wedge\cdots\wedge dx^{i_p}$$

in M to be a Killing form is

$$g^{ji}\nabla_j\nabla_i\omega_{i_1 i_2\cdots i_p} + \frac{1}{p}\sum_{s=1}^{p}K_{i_s}{}^{t}\omega_{i_1\cdots t\cdots i_p} + \frac{1}{p}\sum_{t<s}^{1\cdots p}K_{i_t i_s}{}^{vu}\omega_{i_1\cdots v\cdots u\cdots i_p} = 0 \tag{2.5}$$

and

$$g^{ji}\,\nabla_j\omega_{ii_2\cdots i_p} = 0. \tag{2.6}$$

§3. Curvature and Betti Numbers

Consider the square length $|\omega|^2$ of a p-form ω. We have

$$\tfrac{1}{2}\Delta\,|\omega|^2 = \langle\delta\,\nabla\omega,\omega\rangle - |\nabla\omega|^2,$$

and consequently, integrating over the manifold M, we have

$$\int_M\langle\delta\nabla\omega,\omega\rangle\,dV = \int_M|\nabla\omega|^2\,dV. \tag{3.1}$$

Suppose now that ω is a harmonic p-form, then formula (1.12) holds, that is,

$$(\delta\nabla\overset{*}{\omega})_{i_1 i_2\cdots i_p} = -\sum_{s=1}^{p}K_{i_s}{}^{t}\omega_{i_1\cdots t\cdots i_p} - \sum_{t<s}^{1\cdots p}K_{i_t i_s}{}^{vu}\omega_{i_1\cdots v\cdots u\cdots i_p}.$$

Substituting this into (3.1), we obtain

$$p\int_M F_p(\omega,\omega)\,dV + \int_M|\nabla\omega|^2\,dV = 0, \tag{3.2}$$

where

$$F_p(\omega,\omega) = K_{ji}\omega^{j}{}_{i_2\cdots i_p}\omega^{ii_2\cdots i_p} + \frac{p-1}{2}K_{kjih}\omega^{kj}{}_{i_3\cdots i_p}\omega^{ih i_3\cdots i_p}. \tag{3.3}$$

Thus, if

$$F_p(\omega,\omega)\geq 0,$$

then

$$F_p(\omega, \omega) = 0, \qquad \nabla\omega = 0.$$

If $F_p(\omega, \omega)$ is positive-definite, then $\omega = 0$. Thus, by a theorem of Hodge stated in Section 2 of Chapter 1, we have

THEOREM 3.1. *If, in a compact and orientable Riemannian manifold, the quadratic form $F_p(\omega, \omega)$ is positive-semidefinite, then any harmonic p-form has vanishing covariant derivative. If the quadratic form $F_p(\omega, \omega)$ is positive-definite, then there exists no harmonic p-form other than zero, and consequently*

$$b_p(M) = 0 \qquad (p = 1, 2, \ldots, n-1),$$

$b_p(M)$ being the p-th Betti number of M, (Lichnerowicz [1, 2, 10], Yano [7], Yano and Bochner [1]).

When the Riemannian manifold M is of constant curvature,

$$K_{kjih} = \frac{r}{n(n-1)}(g_{kh}g_{ji} - g_{jh}g_{ki}),$$

and consequently

$$K_{ji} = \frac{r}{n}g_{ji}.$$

Substituting for these curvature components in (3.3), we find that

$$F_p(\omega, \omega) = \frac{n-p}{n(n-1)}r\,|\omega|^2.$$

Thus, $F_p(\omega, \omega)$ being positive-definite for $r > 0$, we have

THEOREM 3.2. *For a compact and orientable Riemannian manifold of positive constant curvature,*

$$b_p(M) = 0 \qquad (p = 1, 2, \ldots, n-1).$$

Assume now that M is conformally flat and $\dim M > 3$. Then the curvature tensor is of the form

$$K_{kjih} = \frac{1}{n-2}(g_{kh}K_{ji} - g_{jh}K_{ki} + K_{kh}g_{ji} - K_{jh}g_{ki}) - \frac{r}{(n-1)(n-2)}(g_{kh}g_{ji} - g_{jh}g_{ki}).$$

Substituting this into (3.3), we find

$$F_p(\omega, \omega) = \frac{n - 2p}{n - 2} K_{ji}\omega^j{}_{i_2\cdots i_p}\omega^{ii_2\cdots i_p} + \frac{p - 1}{(n - 1)(n - 2)} r\,|\omega|^2.$$

Assume that $R(v, v)$ is positive-definite and denote by L the smallest (positive) eigenvalue of the matrix (K_{ji}). Then,

$$R(v, v) \geq Lg(v, v)$$

and

$$r = nL > 0.$$

Hence,

$$F_p(\omega, \omega) \geq \frac{n - 2p}{n - 2} L\,|\omega|^2 + \frac{n(p - 1)}{(n - 1)(n - 2)} L\,|\omega|^2$$
$$= \frac{n - p}{n - 1} L\,|\omega|^2,$$

which is positive-definite. Hence,

THEOREM 3.3. *A compact and orientable conformally flat Riemannian manifold of dimension $n > 3$ with positive-definite Ricci curvature has $b_p = 0$ $(p = 1, 2, \ldots, n - 1)$.*

Assume that the curvature tensor K_{kjih} of M satisfies the inequalities

$$0 < \tfrac{1}{2}k \leq -\frac{K_{kjih}\omega^{kj}\omega^{ih}}{\omega_{ji}\omega^{ji}} \leq k \tag{3.4}$$

for a certain positive number k and for any skew-symmetric tensor field ω of type (0, 2).

Taking two mutually orthogonal unit vectors X and Y and putting

$$\omega^{ji} = Y^jX^i - Y^iX^j,$$

we find, from (3.4),

$$0 < \tfrac{1}{4}k \leq -K_{kjih}Y^kX^jY^iX^h \leq \tfrac{1}{2}k. \tag{3.5}$$

Put $X_1 = X$ and take $n - 1$ unit vectors $X_2, X_3, \cdots, X_n$ which are orthogonal to X_1 and to each other. Then

$$\sum_{s=1}^{n} X_s^jX_s^i = g^{ji}.$$

On the other hand, from (3.5),

$$0 < \tfrac{1}{4}k \leq -K_{kjih}X_s^kX^jX_s^iX^h \leq \tfrac{1}{2}k,$$

from which

$$0 < \frac{n-1}{4} k \leq R(X, X) \leq \frac{n-1}{2} k.$$

Thus,

$$F_p(\omega, \omega) = R_{ji}\omega^j{}_{i_2\cdots i_p}\omega^{ii_2\cdots i_p} + \frac{p-1}{2} K_{kjih}\omega^{kj}{}_{i_3\cdots i_p}\omega^{ihi_3\cdots i_p}$$

$$\geq \frac{n-1}{4} k\,|\omega|^2 - \frac{p-1}{2} k\,|\omega|^2$$

$$= \frac{n-2p+1}{4} k\,|\omega|^2,$$

which is positive-definite for $n \geq 2p$. Hence

THEOREM 3.4. *If, in a compact and orientable Riemannian manifold M, the curvature tensor satisfies the inequalities*

$$0 < \tfrac{1}{2}k \leq -\frac{K_{kjih}\omega^{kj}\omega^{ih}}{\omega_{ji}\omega^{ji}} \leq k$$

for a certain positive constant k and for any bivector ω^{ji}, then

$$b_p(M) = 0 \quad \left(p = 1, 2, \ldots, \left[\frac{n}{2}\right]\right).$$

§4. Curvature and Killing Tensor Fields

Suppose that ω is a Killing p-form, then we have (2.5), that is,

$$(\delta\nabla\omega)_{i_1i_2\cdots i_p} = \frac{1}{p}\sum_{s=1}^{p} K_{i_s}{}^t\omega_{i_1\cdots t\cdots i_p} + \frac{1}{p}\sum_{t<s}^{1\cdots p} K_{i_ti_s}{}^{vu}\omega_{i_1\cdots v\cdots u\cdots i_p}.$$

Substituting this into (2.1), we find

$$\int_M F_p(\omega, \omega)\, dV - p\int_M |\nabla\omega|^2\, dV = 0, \tag{4.1}$$

where $F_p(\omega, \omega)$ is defined by (3.3). Thus,

THEOREM 4.1. *If, in a compact and orientable Riemannian manifold, the quadratic form $F_p(\omega, \omega)$ is negative-definite, then there does not exist a Killing p-form other than zero* (Yano [7]).

When the Riemannian manifold M is of constant curvature we have

$$F_p(\omega, \omega) = \frac{n-p}{n(n-1)} r |\omega|^2.$$

Thus, since $F_p(\omega, \omega)$ is negative-definite for $r < 0$, we have

THEOREM 4.2. *For a compact and orientable Riemannian manifold of negative constant curvature, there is no Killing p-form* ($p = 1, 2, \ldots, n-1$) *other than zero.*

Assume now that M is conformally flat, dim $M > 3$. Then,

$$F_p(\omega, \omega) = \frac{n-2p}{n-2} K_{ji}\omega^j{}_{i_2\cdots i_p}\omega^{ii_2\cdots i_p} + \frac{p-1}{(n-1)(n-2)} r |\omega|^2.$$

Assume moreover that $R(v, v)$ is negative-definite and denote by $-M$ the largest (negative) eigenvalue of the matrix (K_{ji}). Then

$$R(v, v) \leq -Mg(v, v)$$

and

$$r = -nM < 0.$$

Hence, for $n \geq 2p$,

$$\begin{aligned} F_p(\omega, \omega) &\leq -\frac{n-2p}{n-2} M |\omega|^2 - \frac{n(p-1)}{(n-1)(n-2)} M |\omega|^2 \\ &= -\frac{n-p}{n-1} M |\omega|^2, \end{aligned}$$

which is negative-definite. Hence,

THEOREM 4.3. *For a compact and orientable conformally flat Riemannian manifold M of dimension n > 3 with negative-definite Ricci curvature, there exists no Killing tensor field of order p other than zero* ($p = 1, 2, \ldots, [n/2]$).

We next assume that the curvature tensor K_{kjih} of M satisfies the inequalities

$$-k \leq -\frac{K_{kjih}\omega^{kj}\omega^{ih}}{\omega_{ji}\omega^{ji}} \leq -\tfrac{1}{2}k < 0 \tag{4.2}$$

for a certain negative number $-k$ and for any skew-symmetric tensor field ω of type $(0, 2)$.

Then,

$$F_p(\omega, \omega) \leq -\frac{n-2p+1}{4} k\,|\omega|^2,$$

which is negative-definite for $n \geq 2p$. Hence,

THEOREM 4.4. *If, in a compact and orientable Riemannian manifold, the curvature tensor satisfies the inequalities*

$$-k \leq -\frac{K_{kjih}\omega^{kj}\omega^{ih}}{\omega_{ji}\omega^{ji}} \leq -\tfrac{1}{2}k < 0$$

for a certain negative constant $-k$ *and for any bivector* ω^{ji}, *then there exists no Killing tensor field of order* p *other than zero* ($p = 1, 2, \ldots, [n/2]$).

PROBLEMS

1. Prove that if a harmonic form is coderived, then it is identically zero.
2. In a Riemannian manifold, the Weyl projective curvature tensor is given by

$$P_{kjih} = K_{kjih} - \frac{1}{n-1}(K_{kh}g_{ji} - K_{jh}g_{ki}).$$

If the tensor P_{kjih} of a Riemannian manifold of dimension $n > 2$ vanishes, the manifold is projectively flat and consequently of constant curvature, so that the tensor represents the deviation of the manifold from projective flatness. In order to measure the deviation from projective flatness, we introduce the quantity

$$P = \sup_\omega \frac{|P_{kjih}\omega^{kj}\omega^{ih}|}{\omega_{ji}\omega^{ji}} \qquad (\omega_{ji} = -\omega_{ij}).$$

We assume that the Ricci curvature $K_{ji}\omega^j\omega^i$ is positive-definite and denote by L the smallest positive eigenvalue of this quadratic form. Then,

$$K_{ji}\omega^j\omega^i \geq L\omega_i\omega^i$$

and

$$r = g^{ji}K_{ji} \geq nL > 0.$$

Prove the following

THEOREM. *In a compact and orientable Riemannian manifold* M *of positive Ricci curvature, if*

$$\frac{n-p}{n-1}L > \frac{p-1}{2}P,$$

then there exists no harmonic form of order p *other than the zero form, and consequently* $b_p(M) = 0$ ($p = 1, 2, \cdots, n-1$) (Bochner [5], Goldberg [4], Yano [7], Yano and Bochner [1]).

3. The quantity P is defined as in Problem 2. We assume that the Ricci curvature $K_{ji}\omega^j\omega^i$ is negative-definite and denote by $-M$ the largest negative eigenvalue of this form. Then,

$$K_{ji}\omega^j\omega^i \leq -M\omega_i\omega^i$$

and

$$r = g^{ji}K_{ji} \leq -nM < 0.$$

Prove the following

THEOREM. *In a compact and orientable Riemannian manifold M of negative-definite Ricci curvature, if*

$$\frac{n-p}{n-1}M > \frac{p-1}{2}P,$$

then there exists no Killing tensor field of order p other than the zero tensor ($p = 1, 2, \ldots, n-1$).

4. In a Riemannian manifold, the concircular curvature tensor is given by

$$Z_{kjih} = K_{kjih} - \frac{1}{n(n-1)}r(g_{kh}g_{ji} - g_{jh}g_{ki}).$$

If the tensor Z_{kjih} of a Riemannian manifold of dimension $n > 2$ vanishes, the manifold is said to be *concircularly flat* and consequently of constant curvature; so the concircular curvature tensor represents deviation of the manifold from constant curvature. In order to measure this deviation, we introduce the norm

$$Z = \sup_\omega \frac{|Z_{kjih}\omega^{kj}\omega^{ih}|}{\omega_{ji}\omega^{ji}} \qquad (\omega_{ji} = -\omega_{ij}).$$

The quantities L and M are defined as in Problems 2 and 3. Prove

THEOREM. *In a compact and orientable Riemannian manifold M of positive Ricci curvature, if*

$$L - \frac{p-1}{n(n-1)}r > \frac{p-1}{2}Z,$$

then $b_p(M) = 0, p = 1, 2, \ldots, n-1$, *and for negative Ricci curvature, if*

$$M + \frac{p-1}{n(n-1)}r > \frac{p-1}{2}Z,$$

then there exists no Killing tensor of order p other than zero, $p = 1,2, \ldots, n-1$ (Yano [7], Yano and Bochner [1]).

5. In a Riemannian manifold, the covariant components of the Weyl conformal curvature tensor are given by

$$C_{kjih} = K_{kjih} - \frac{1}{n-2}(g_{kh}K_{ji} - g_{jh}K_{ki} + K_{kh}g_{ji} - K_{jh}g_{ki})$$
$$+ \frac{1}{(n-1)(n-2)}(g_{kh}g_{ji} - g_{jh}g_{ki}).$$

If the Weyl conformal curvature tensor vanishes and $n > 3$, the manifold is conformally flat; so this tensor represents the deviation of the manifold from conformal flatness. In order to measure deviation from conformal flatness, we introduce the norm

$$C = \sup_{\omega} \frac{|C_{kjih}\omega^{kj}\omega^{ih}|}{\omega_{ji}\omega^{ji}} \qquad (\omega_{ji} = -\omega_{ij}).$$

The quantities L and M are defined as in Problems 2 and 3. Prove

THEOREM. *In a compact and orientable Riemannian manifold M of positive Ricci curvature, if*

$$\frac{n-2p}{n-2}L + \frac{p-1}{(n-1)(n-2)}r > \frac{p-1}{2}C, \qquad n > 3,$$

then $b_p(M) = 0, p = 1, 2, \ldots, n-1$, *and for negative Ricci curvature, if*

$$\frac{n-2p}{n-2}M - \frac{p-1}{(n-1)(n-2)}r > \frac{p-1}{2}C, \qquad n > 3,$$

then, there is no Killing tensor field of order p other than zero, $p = 1, 2, \ldots, [n/2]$ (Bochner [5], Goldberg [4], Mogi [1], Yano [7], Yano and Bochner [1]).

CHAPTER

5

Hypersurfaces of Riemannian Manifolds

In Sections 1 and 2 of this chapter, some integral formulas for hypersurfaces of a Euclidean space are established and applications of these formulas are given.

In the remaining part of the chapter, we recall some of the fundamental concepts and formulas in the theory of hypersurfaces in Riemannian manifolds.

§1. Fundamental Formulas for Hypersurfaces of a Euclidean Space

We consider an orientable differentiable hypersurface V covered by a system of coordinate neighborhoods $\{U; x^h\}$ and imbedded differentiably in an $(n + 1)$-dimensional Euclidean space E referred to a rectangular coordinate system. If we denote by X the position vector from the origin O to a point P of the hypersurface V, then the hypersurface is represented by

$$X = X(x^h). \tag{1.1}$$

The

$$X_i = \partial_i X, \qquad \partial_i = \partial/\partial x^i, \tag{1.2}$$

are n linearly independent vector fields tangent to the hypersurface.

Suppose that the local coordinates x^h are chosen in such a way that the vectors $X_1, X_2, \ldots, X_n$ give the positive orientation of V. Then

$$X_j \cdot X_i = g_{ji} \tag{1.3}$$

give the components of the metric tensor of V with respect to the system of coordinate neighborhoods $\{U, x^h\}$, where the "$\cdot$" denotes the inner product of vectors in E. We choose the unit normal vector N to V in E in such a way that the vectors $N, X_1, \ldots, X_n$ defined along the hypersurface V give the positive orientation of E. Then,

$$X_i \cdot N = 0, \qquad N \cdot N = 1. \\ |N, X_1, \ldots, X_n| > 0, \tag{1.4}$$

where $|N, X_1, \ldots, X_n|$ denotes the determinant formed with the components of $N, X_1, \ldots, X_n$ in this order.

We denote by ∇_i the operator of covariant differentiation with respect to the Christoffel symbols $\left\{ {h \atop ji} \right\}$ formed with the g_{ji}. Then the Gauss equations of the hypersurface are

$$\nabla_j X_i \equiv \partial_j X_i - \left\{ {h \atop ji} \right\} X_h = h_{ji} N, \tag{1.5}$$

where the h_{ji} are the components of the *second fundamental tensor* of the hypersurface. The equations of Weingarten are

$$\nabla_j N \equiv \partial_j N = -h_j{}^t X_t, \tag{1.6}$$

where

$$h_j{}^t = h_{ji} g^{it},$$

the g^{it} being the contravariant components of the metric tensor.

Employing the Ricci identities

$$\nabla_k \nabla_j X_i - \nabla_j \nabla_k X_i = -K_{kji}{}^h X_h$$

and

$$\nabla_k \nabla_j N - \nabla_j \nabla_k N = 0,$$

where the $K_{kji}{}^h$ are the components of the curvature tensor of g_{ji}, we obtain the equations of Gauss,

$$K_{kji}{}^t = h_k{}^t h_{ji} - h_j{}^t h_{ki}, \tag{1.7}$$

and those of Codazzi,

$$\nabla_k h_{ji} - \nabla_j h_{ki} = 0. \tag{1.8}$$

The principal curvature of the hypersurface V are the roots of the determinantal equation

$$|h_i{}^h - k\delta_i^h| = 0. \tag{1.9}$$

We denote them by $k_1, k_2, \ldots, k_n$ and put

$$\begin{aligned} s_0 &= 1, \qquad s_l = \sum_{i_1<i_2<\cdots<i_l} k_{i_1}k_{i_2}\cdots k_{i_l}, \\ p_0 &= n, \qquad p_l = \sum_i (k_i)^l \quad (l = 1, 2, \ldots, n). \end{aligned} \tag{1.10}$$

From (1.9) and (1.10) we see that

$$p_l = h_{i_2}{}^{i_1}h_{i_3}{}^{i_2}\cdots h_{i_l}{}^{i_{l-1}}h_{i_1}{}^{i_l}. \tag{1.11}$$

It is well-known that $s_1, \ldots, s_n$ and $p_1, \ldots, p_n$ are related by Newton's formulas:

$$\begin{aligned} &p_1 - s_1 = 0, \\ &p_2 - s_1p_1 + 2s_2 = 0, \\ &p_3 - s_1p_2 + s_2p_1 - 3s_3 = 0, \\ &\cdots \\ &p_n - s_1p_{n-1} + s_2p_{n-2} - \cdots - (-1)^n s_{n-1}p_1 + (-1)^n n s_n = 0. \end{aligned} \tag{1.12}$$

Expressing the s_l in terms of $p_1, p_2, \cdots, p_l$, we obtain

$$\begin{aligned} s_1 &= p_1, \\ s_2 &= \frac{1}{2!}(-p_2 + p_1^2), \\ s_3 &= \frac{1}{3!}(2p_3 - 3p_1p_2 + p_1^3), \\ s_4 &= \frac{1}{4!}(-6p_4 + 8p_1p_3 - 6p_1^2p_2 + 3p_2^2 - p_1^4), \\ &\cdots \\ s_l &= \sum_{\substack{t_1+2t_2+\cdots+lt_l=l \\ 0\le t_i}} \frac{(-1)^{t_1+t_2+\cdots+t_l+l}}{(t_1!)\cdots(t_l!)2^{t_2}\cdots l^{t_l}} p_1^{t_1}p_2^{t_2}\cdots p_l^{t_l}, \\ &\cdots \\ s_n &= \sum_{\substack{t_1+2t_2+\cdots+nt_n=n \\ 0\le t_i}} \frac{(-1)^{t_1+\cdots+t_n+n}}{(t_1!)\cdots(t_n!)2^{t_2}\cdots n^{t_n}} p_1^{t_1}p_2^{t_2}\cdots p_n^{t_n}. \end{aligned} \tag{1.13}$$

Setting

$$h_{(l)i}{}^h = h_{i_1}{}^h h_{i_2}{}^{i_1}\cdots h_i{}^{i_{l-1}}, \tag{1.14}$$

$$\begin{aligned} z_{(l)}{}^h &= h_{(l)i}{}^h z^i \\ &= h_{i_1}{}^h h_{i_2}{}^{i_1}\cdots h_i{}^{i_{l-1}} z^i \quad (l = 1, 2, \ldots, n), \end{aligned} \tag{1.15}$$

where $z_{(0)}{}^h = z^h$ is a vector field of V and

$$\binom{n}{l} M_l = s_l, \tag{1.16}$$

where the $\binom{n}{l}$ are binomial coefficients, M_l is the lth *mean curvature* of V in E. From (1.15),

$$z_{(l)}{}^h = h_i{}^h z_{(l-1)}{}^i. \tag{1.17}$$

Since g_{ji} is positive-definite and h_{ji} is symmetric in j and i, we can assume that, at a fixed point of V,

$$(g_{ji}) = \begin{pmatrix} 1 & 0 & \cdots & 0 \\ 0 & 1 & \cdots & 0 \\ \cdot & & & \cdot \\ \cdot & & & \cdot \\ \cdot & & & \cdot \\ 0 & 0 & \cdots & 1 \end{pmatrix}, \qquad (h_{ji}) = \begin{pmatrix} k_1 & 0 & \cdots & 0 \\ 0 & k_2 & \cdots & 0 \\ \cdot & & & \cdot \\ \cdot & & & \cdot \\ \cdot & & & \cdot \\ 0 & 0 & \cdots & k_n \end{pmatrix}.$$

Consequently,

$$(h_i{}^r) = \begin{pmatrix} k_1 & 0 & \cdots & 0 \\ 0 & k_2 & \cdots & 0 \\ \cdot & & & \cdot \\ \cdot & & & \cdot \\ \cdot & & & \cdot \\ 0 & 0 & \cdots & k_n \end{pmatrix}$$

and

$$h_{(l)i}{}^r = \begin{pmatrix} k_1^l & 0 & \cdots & 0 \\ 0 & k_2{}^l & \cdots & 0 \\ \cdot & & & \cdot \\ \cdot & & & \cdot \\ \cdot & & & \cdot \\ 0 & 0 & \cdots & k_n{}^l \end{pmatrix}.$$

Now, the $k_1, k_2, \ldots, k_n$ satisfy the equation

$$t^n - s_1 t^{n-1} + s_2 t^{n-2} - \cdots + (-1)^{n-1} s_{n-1} t + (-1)^n s_n = 0,$$

and consequently

$$h_{(n)i}{}^h - s_1 h_{(n-1)i}{}^h + s_2 h_{(n-2)i}{}^h - \cdots + (-1)^{n-1} s_{n-1} h_i{}^h + (-1)^n s_n \delta_i^h = 0,$$

or

$$h_{(n)ji} - s_1 h_{(n-1)ji} + s_2 h_{(n-2)ji} - \cdots + (-1)^{n-1} s_{n-1} h_{ji} + (-1)^n s_n g_{ji} = 0. \tag{1.18}$$

In the sequel we will need the expression for $\nabla_i z_{(l)}{}^i$. We have

$$\begin{aligned}\nabla_i z_{(l)}{}^i &= \nabla_i(h_{i_1}{}^i h_{i_2}{}^{i_1} \cdots h_{i_{l-1}}{}^{i_{l-2}} h_{i_l}{}^{i_{l-1}} z^{i_l}) \\ &= (\nabla_{i_1} h_i^i) h_{i_2}^{i_1} \cdots h_{i_{l-1}}{}^{i_{l-2}} h_{i_l}{}^{i_{l-1}} z^{i_l} \\ &\quad + h_{i_1}{}^i (\nabla_{i_2} h_i{}^{i_1}) \cdots h_{i_{l-1}}{}^{i_{l-2}} h_{i_l}{}^{i_{l-1}} z^{i_l} \\ &\quad \cdots \\ &\quad + h_{i_1}{}^i h_{i_2}{}^{i_1} \cdots (\nabla_{i_{l-1}} h_i{}^{i_{l-2}}) h_{i_l}{}^{i_{l-1}} z^{i_l} \\ &\quad + h_{i_1}{}^i h_{i_2}{}^{i_1} \cdots h_{i_{l-1}}{}^{i_{l-2}} (\nabla_{i_l} h_i^{i_{l-1}}) z^{i_l} \\ &\quad + h_{i_1}{}^i h_{i_2}{}^{i_1} \cdots h_{i_{l-1}}{}^{i_{l-2}} h_{i_l}{}^{i_{l-1}} (\nabla_i z^{i_l})\end{aligned}$$

by virtue of the equations of Codazzi (see Eq. (1.8)), and consequently

$$\begin{aligned}\nabla_i z_{(l)}{}^i &= (\nabla_i p_1) z_{(l-1)}{}^i + \tfrac{1}{2}(\nabla_i p_2) z_{(l-2)}{}^i + \cdots \\ &\quad + \frac{1}{l-1}(\nabla_i p_{l-1}) z_{(1)}{}^i + \frac{1}{l}(\nabla_i p_l) z^i + h_{(l)h}{}^i \nabla_i z^h. \qquad (1.19)\end{aligned}$$

§2. Integral Formulas for Hypersurfaces of a Euclidean Space

We consider a compact and orientable hypersurface V of an $(n + 1)$-dimensional Euclidean space E and put

$$z_i = X \cdot X_i, \qquad \alpha = X \cdot N. \qquad (2.1)$$

$$\nabla_j z_i = g_{ji} + \alpha h_{ji}, \qquad (2.2)$$

$$\nabla_j \alpha = -h_{ji} z^i \qquad (2.3)$$

by virtue of the equations of Gauss and Weingarten and consequently, from (1.19),

$$\begin{aligned}\nabla_i z_{(l)}{}^i &= (\nabla_i p_1) z_{(l-1)}{}^i + \tfrac{1}{2}(\nabla_i p_2) z_{(l-2)}{}^i + \cdots \\ &\quad + \frac{1}{l-1}(\nabla_i p_{l-1}) z_{(1)}{}^i + \frac{1}{l}(\nabla_i p_l) z^i + p_l + \alpha p_{l+1}. \qquad (2.4)\end{aligned}$$

For $l = 0, 1, 2$, Eq. (2.4) gives

$$\nabla_i z^i = n + \alpha p_1, \qquad (2.5)$$

$$\nabla_i z_{(1)}{}^i = (\nabla_i p_1) z^i + p_1 + \alpha p_2, \qquad (2.6)$$

$$\nabla_i z_{(2)}{}^i = (\nabla_i p_1) z_{(1)}{}^i + \tfrac{1}{2}(\nabla_i p_2) z^i + p_2 + \alpha p_3, \qquad (2.7)$$

respectively.

Thus, integrating (2.5) over V, we find

$$\int_V (n + \alpha p_1)\, dV = 0, \tag{2.8}$$

dV being the volume element of V. Substituting

$$p_1 = s_1 = nM_1$$

in Eq. (2.8),

$$\int_V (1 + \alpha M_1)\, dV = 0, \tag{2.9}$$

which shows that M_1 cannot vanish identically on V. Integrating (2.6) over V, we find

$$\int_V \{(\nabla_i p_1)z^i + p_1 + \alpha p_2\}\, dV = 0. \tag{2.10}$$

On the other hand,

$$\nabla_i(p_1 z^i) = (\nabla_i p_1)z^i + p_1(n + \alpha p_1)$$

by virtue of (2.5), from which, integrating over V,

$$\int_V \{(\nabla_i p_1)z^i + p_1(n + \alpha p_1)\}\, dV = 0. \tag{2.11}$$

Thus, subtracting (2.10) from (2.11), we obtain

$$\int_V \{(n - 1)p_1 + \alpha(p_1^2 - p_2)\}\, dV = 0, \tag{2.12}$$

or

$$\int_V \{(n - 1)s_1 + 2\alpha s_2\}\, dV = 0 \tag{2.13}$$

by virtue of (1.13), or, again,

$$\int_V (M_1 + \alpha M_2)\, dV = 0, \tag{2.14}$$

by virtue of (1.16).

Using the integral formulas (2.9) and (2.14), we obtain

THEOREM 2.1. *If a compact and orientable hypersurface V of an $(n + 1)$-dimensional Euclidean space satisfies $M_1 =$* const *and the function α has constant sign, then V is totally umbilical and so is a sphere* (Liebmann [1], Süss [1]).

Proof. Multiplying (2.9) by M_1 and then subtracting (2.14), we find

$$\int_V \alpha(M_1^2 - M_2)\, dV = 0.$$

On the other hand,

$$M_1^2 - M_2 = \frac{1}{n^2(n-1)} \sum_{j<i} (k_j - k_i)^2 \geq 0.$$

Consequently, if α has constant sign,

$$M_1^2 - M_2 = 0;$$

that is,

$$k_1 = k_2 = \cdots = k_n.$$

It follows that V is a compact and orientable totally umbilical hypersurface in a Euclidean space, and hence a sphere.

Integrating (2.7) over V, we obtain

$$\int_V \{(\nabla_i p_1) z_{(1)}{}^i + \tfrac{1}{2}(\nabla_i p_2) z^i + p_2 + \alpha p_3\}\, dV = 0. \tag{2.15}$$

On the other hand,

$$\begin{aligned} \nabla_i(p_1 z_{(1)}{}^i) &= (\nabla_i p_1) z_{(1)}{}^i + p_1 \nabla_i z_{(1)}{}^i \\ &= (\nabla_i p_1) z_{(1)}{}^i + p_1\{(\nabla_i p_1) z^i + p_1 + \alpha p_2\} \end{aligned}$$

by virtue of (2.6), from which, integrating over V,

$$\int_V \{(\nabla_i p_1) z_{(1)}{}^i + \tfrac{1}{2}(\nabla_i p_1^2) z^i + p_1(p_1 + \alpha p_2)\}\, dV = 0. \tag{2.16}$$

We also have

$$\begin{aligned} \nabla_i\{\tfrac{1}{2}(p_1^2 - p_2) z^i\} &= \tfrac{1}{2}\{\nabla_i(p_1^2 - p_2)\} z^i + \tfrac{1}{2}(p_1^2 - p_2)\nabla_i z^i \\ &= \tfrac{1}{2}\{\nabla_i(p_1^2 - p_2)\} z^i + \tfrac{1}{2}(p_1^2 - p_2)(n + \alpha p_1) \end{aligned}$$

by virtue of (2.5), from which, integrating over V,

$$\int_V [\tfrac{1}{2}\{\nabla_i(p_1^2 - p_2)\} z^i + \tfrac{1}{2}(p_1^2 - p_2)(n + \alpha p_1)]\, dV = 0. \tag{2.17}$$

Subtracting (2.15) and (2.17) from (2.16) yields

$$\int_V \{(n-2)(p_1^2 - p_2) + \alpha(p_1^3 - 3p_1 p_2 + 2p_3)\}\, dV = 0. \tag{2.18}$$

or

$$\int_V \{(2!)(n-2) s_2 + (3!)\, \alpha s_3\}\, dV = 0, \tag{2.19}$$

by virtue of (1.13), or, again,

$$\int_V (M_2 + \alpha M_3)\, dV = 0, \tag{2.20}$$

by virtue of (1.16).

To obtain integral formulas for the most general case, we compute

$$\begin{aligned}\nabla_i(z_{(l)}{}^i) = (\nabla_i p_1) z_{(l-1)}{}^i + \tfrac{1}{2}(\nabla_i p_2) z_{(l-2)}{}^i + \cdots \\ + \frac{1}{l-1}(\nabla_i p_{l-1}) z_{(1)}{}^i + \frac{1}{l}(\nabla_i p_l) z^i + p_l + \alpha p_{l+1},\end{aligned} \tag{2.21}$$

$$\begin{aligned}\nabla_i(p_1 z_{(l-1)}{}^i) = (\nabla_i p_1) z_{(l-1)}{}^i + p_1\{(\nabla_i p_1) z_{(l-2)}{}^i + \tfrac{1}{2}(\nabla_i p_2) z_{(l-3)}{}^i + \cdots \\ + \frac{1}{l-2}(\nabla_i p_{l-2}) z_{(1)}{}^i + \frac{1}{l-1}(\nabla_i p_{l-1}) z^i + p_{l-1} + \alpha p_l\},\end{aligned} \tag{2.22}$$

$$\begin{aligned}&\nabla_i\{\tfrac{1}{2}(p_1^2 - p_2) z_{(l-2)}{}^i\} \\ &= \frac{1}{2}\Big\{\{\nabla_i(p_1^2 - p_2)\} z_{(l-2)}{}^i + (p_1^2 - p_2)\{(\nabla_i p_1) z_{(l-3)}{}^i + \tfrac{1}{2}(\nabla_i p_2) z_{(l-4)}{}^i + \cdots \\ &\quad + \frac{1}{l-3}(\nabla_i p_{l-3}) z_{(1)}{}^i + \frac{1}{l-2}(\nabla_i p_{l-2}) z^i + p_{l-2} + \alpha p_{l-1}\}\Big\}.\end{aligned} \tag{2.23}$$

$$\begin{aligned}&\nabla_i\left\{\frac{1}{3!}(p_1^3 - 3p_1p_2 + 2p_3) z_{(l-3)}{}^i\right\} \\ &= \frac{1}{3!}\Big[\{\nabla_i(p_1^3 - 3p_1p_2 + 2p_3)\} z_{(l-3)}{}^i + (p_1^3 - 3p_1p_2 + 2p_3) \\ &\quad \times \Big\{(\nabla_i p_1) z_{(l-4)}{}^i + \tfrac{1}{2}(\nabla_i p_2) z_{(l-5)}{}^i + \cdots \\ &\quad + \frac{1}{l-4}(\nabla_i p_{l-4}) z_{(1)}{}^i + \frac{1}{l-3}(\nabla_i p_{l-3}) z^i + p_{l-3} + \alpha p_{l-2}\Big\}\Big], \\ &\qquad \cdots\end{aligned} \tag{2.24}$$

$$\begin{aligned}&\nabla_i\left\{\sum_{\substack{t_1+2t_2+\cdots+lt_l=l \\ 0\le t_i}} \frac{(-1)^{t_1+\cdots+t_l+l}}{(t_1!)\cdots(t_l!)2^{t_2}\cdots n^{t_l}} p_1^{t_1}\cdots p_l^{t_l} z^i\right\} \\ &= \sum_{\substack{t_1+2t_2+\cdots+lt_l=l \\ 0\le t_i}} \frac{(-1)^{t_1+\cdots+t_l+l}}{(t_1!)\cdots(t_l!)2^{t_2}\cdots n^{t_l}} \\ &\quad \times \{(\nabla_i p_1^{t_1}\cdots p_l^{t_l}) z^i + p_1^{t_1}\cdots p_l^{t_l}(n + \alpha p_1)\}.\end{aligned} \tag{2.25}$$

Integrating $-(2.21) + (2.22) - (2.23) + (2.24) - \cdots + (-1)^l(2.25)$, we find

$$\int_V \Big\{(p_l + \alpha p_{l+1}) - s_1(p_{l-1} + \alpha p_l) + s_2(p_{l-2} + \alpha p_{l-3})$$

$$- s_3(p_{l-3} + \alpha p_{l-2}) + \cdots + (-1)^l s_l(n + \alpha p_1)\Big\}\, dV = 0, \quad (2.26)$$

or, by (1.12),

$$\int_V \{(n - l)s_l + \alpha(l + 1)s_{l+1}\}\, dV = 0, \quad (2.27)$$

or again, by (1.16),

$$\int_V (M_l + \alpha M_{l+1})\, dV = 0. \quad (2.28)$$

Thus,

$$\begin{aligned} &\int_V (1 + \alpha M_1)\, dV = 0, \\ &\int_V (M_1 + \alpha M_2)\, dV = 0, \\ &\int_V (M_2 + \alpha M_3)\, dV = 0, \\ &\qquad \cdots \\ &\int_V (M_l + \alpha M_{l+1})\, dV = 0, \\ &\qquad \cdots \\ &\int_V (M_{n-1} + \alpha M_n)\, dV = 0. \end{aligned} \quad (2.29)$$

Using these integral formulas, we can prove

THEOREM 2.2. *If a closed convex hypersurface V of an $(n + 1)$-dimensional Euclidean space satisfies $M_l =$ const for a fixed l, then V is totally umbilical, and hence a sphere.*

Proof. From

$$\int_V (1 + \alpha M_1)\, dV = 0$$

and

$$\int_V (M_l + \alpha M_{l+1})\, dV = 0,$$

we have

$$\int_V \alpha(M_1 M_l - M_{l+1})\, dV = 0. \tag{2.30}$$

But

$$M_1 M_l - M_{l+1} = \frac{l!\,(n-l-1)!}{nn!} \sum_{i_1<i_2<i_3<\cdots<i_{l+1}} (k_{i_1} - k_{i_2})^2 k_{i_3} \cdots k_{i_{l+1}}. \tag{2.31}$$

Since V is convex, all the principal curvatures are positive; so we can choose the position of the origin O in such a way that α has a definite sign. Thus

$$\alpha(M_1 M_l - M_{l+1})$$

has constant sign and consequently, from (2.30),

$$M_1 M_l - M_{l+1} = 0,$$

from which

$$k_1 = k_2 = \cdots = k_n.$$

Thus V is totally umbilical, and hence is a sphere.

As to the generalization of the results stated in this section to hypersurfaces of a Riemannian manifold, see Tani [1].

§3. Hypersurfaces of Riemannian Manifolds

Let M be an n-dimensional differentiable manifold of class C^∞ and P an $(n-1)$-dimensional differentiable manifold of class C^∞ imbedded in M with imbedding map

$$i\colon\ P \to M. \tag{3.1}$$

We call the image $i(P)$ a *hypersurface* in M and identify it with the manifold P. If the manifold M is covered by a system of coordinate neighborhoods $\{U; x^h\}$ and P is covered by a system of coordinate neighborhoods $\{U'; u^a\}$, where here and in the sequel the indices $h, i, j, \ldots$ run over the range $1, 2, \ldots, n$ and $a, b, c, \ldots, f$ run over the range $1, 2, \ldots, n-1$, then the imbedding $i\colon\ P \to M$ can be represented locally by

$$x^h = x^h(u^a). \tag{3.2}$$

The differential di of the map i will be denoted by B, so that to a vector X' in P there corresponds a vector BX' in M. Thus, if X' has the local expression $X' = X'^a\, \partial_a$, where $\partial_a = \partial/\partial u^a$, then BX' has the local expression $BX' = B_a{}^i X^a \partial_i$, where $\partial_i = \partial/\partial x^i$ and

$$B_a{}^i = \frac{\partial x^i}{\partial u^a}. \tag{3.3}$$

Let X' be a vector field on P. Then a vector field X defined on M is called an *extension* of the vector field BX' on $i(P)$ if its restriction to $i(P)$ satisfies

$$X|_{i(P)} = BX'. \tag{3.4}$$

We prove

PROPOSITION 3.1. *Let X' and Y' be two vector fields on P, and X and Y extensions of BX' and BY', respectively. Then $[X, Y]|_{i(P)}$ is independent of the extensions, and*

$$[BX', BY'] = B[X', Y'], \tag{3.5}$$

where

$$[BX', BY'] = [X, Y]|_{i(P)}. \tag{3.6}$$

Proof. Denoting by $X^h(x)$ and $Y^h(x)$ the extensions of $B_a{}^h X'^a(u)$ and $B_a{}^h Y'^a(u)$, respectively, we have

$$X^h(x(u)) = B_a{}^h X'^a(u), \qquad Y^h(x(u)) = B_a{}^h Y'^a(u).$$

Consequently, $[X, Y]|_{i(P)}$ has the components

$$\begin{aligned}
[X, Y]^h|_{x=x(u)} &= (X^i \partial_i Y^h - Y^i \partial_i X^h)|_{x=x(u)} \\
&= (B_a{}^i X'^a \partial_i Y^h - B_a{}^i Y'^a \partial_i X^h)|_{x=x(u)} \\
&= X'^a \partial_a (B_b{}^h Y'^b) - Y'^a \partial_a (B_b{}^h X'^b) \\
&= B_b{}^h (X'^a \partial_a Y'^b - Y'^a \partial_a X'^b) \\
&= B_b{}^h [X', Y']^b.
\end{aligned}$$

Thus $[X, Y]|_{i(P)}$ does not depend on the extensions X and Y of BX' and BY' and is equal to $B[X', Y']$.

Suppose that the manifold M is a Riemannian manifold with Riemannian metric g; then the hypersurface P is also a Riemannian manifold with the Riemannian metric

$$g'(X', Y') = g(BX', BY'). \tag{3.7}$$

If the manifolds M and P are both orientable, we can choose a vector field N of M defined along P such that

$$g(BX', N) = 0, \tag{3.8}$$

$$g(N, N) = 1 \tag{3.9}$$

and

$$\sqrt{\det(g)} \cdot \det(N, B) = \sqrt{\det(g')},$$

where "det" denotes the determinant.

§4. The Induced Connection

We denote by ∇ the Riemannian connection in M with respect to g. The Riemannian connection ∇ has no torsion, that is,

$$\nabla_X Y - \nabla_Y X - [X, Y] = 0, \tag{4.1}$$

and is metric, that is,

$$\nabla_X(g(Y, Z)) = g(\nabla_X Y, Z) + g(Y, \nabla_X Z). \tag{4.2}$$

PROPOSITION 4.1. *Let BX' and BY' be two vector fields tangent to the hypersurface $i(P)$. Then $(\nabla_X Y)|_{i(P)}$, where X and Y are extensions of BX' and BY', respectively, does not depend on the extension. Denoting this by $\nabla_{BX'}BY'$,*

$$\nabla_{BX'}BY' = B(\nabla'_{X'}Y') + h(X', Y')N,$$

where ∇' is an affine connection defined on the hypersurface and h is a tensor field of type $(0, 2)$ on the hypersurface.

Proof. Denoting by $\left\{ {h \atop ji} \right\}$ the components of the Riemannian connection ∇ and using the same notation as in the previous section, we have, as components of $\nabla_X Y$,

$$X^j\left(\partial_j Y^h + \left\{ {h \atop ji} \right\} Y^i\right).$$

This gives

$$X^j\left(\partial_j Y^h + \left\{ {h \atop ji} \right\} Y^i\right)\Big|_{x=x(u)} = B_c^{\ j} X'^c\left(\partial_j Y^h + \left\{ {h \atop ji} \right\} B_b^{\ i} Y'^b\right)\Big|_{x=x(u)}$$
$$= X'^c\left(\partial_c(B_b^{\ h} Y'^b) + \left\{ {h \atop ji} \right\} B_c^{\ j} B_b^{\ i} Y'^b\right),$$

so $\nabla_X Y$ does not depend on the extension.

Since $\nabla_{BX'}BY'$ is a vector field defined along the hypersurface, it can be expressed in the form

$$\nabla_{BX'}BY' = BV' + \lambda' N$$

for some vector field V' and function λ' on the hypersurface depending on

X' and Y'. Replacing X' and Y' by $\alpha X'$ and $\beta Y'$, respectively, α and β being functions on the hypersurface, we have

$$\begin{aligned}\nabla_{\alpha BX'}(\beta BY') &= \alpha\nabla_{BX'}(\beta BY')\\ &= \alpha\{(\nabla_{BX'}\beta)BY' + \beta(\nabla_{BX'}BY')\}\\ &= \alpha(X'\beta)BY' + \alpha\beta(BV' + \lambda' N)\\ &= B\{\alpha(X'\beta)Y' + \alpha\beta V'\} + (\alpha\beta)\lambda' N.\end{aligned}$$

Thus, by setting

$$\nabla_{\alpha BX'}(\beta BY') = B\bar{V}' + \bar{\lambda}' N,$$

we have

$$\bar{V}' = \alpha(X'\beta)Y' + (\alpha\beta)V'.$$

and

$$\bar{\lambda}' = (\alpha\beta)\lambda'.$$

The first equation shows that there exists an affine connection ∇' of the hypersurface such that

$$\bar{V}' = \nabla'_{X'}\, Y',$$

and the second shows that λ' is bilinear in X' and Y'; that is,

$$\lambda = h(X', Y').$$

Thus,

$$\nabla_{BX'}BY' = B(\nabla'_{X'}Y') + h(X', Y')N.$$

The affine connection ∇' is called the *connection induced on the hypersurface* by ∇ with respect to the normal N, and h is the *second fundamental tensor* of the hypersurface with respect to N.

PROPOSITION 4.2. *If the connection ∇ of M has no torsion, then the induced connection ∇' has no torsion and the second fundamental tensor h is symmetric.*

Proof. Since ∇ has no torsion, we have

$$\begin{aligned}0 &= \nabla_{BX'}BY' - \nabla_{BY'}BX' - [BX', BY']\\ &= B\nabla'_{X'}Y' + h(X', Y')N - B\nabla'_{Y'}X' - h(Y', X')N - B[X', Y']\\ &= B(\nabla'_{X'}Y' - \nabla'_{Y'}X' - [X', Y']) + (h(X', Y') - h(Y', X'))N,\end{aligned}$$

from which, B and N being linearly independent,

$$\begin{aligned}\nabla'_{X'}Y' - \nabla'_{Y'}X' - [X', Y'] &= 0,\\ h(X', Y') &= h(Y', X').\end{aligned}$$

PROPOSITION 4.3. *If ∇ is the Riemannian connection of a Riemannian manifold, then the connection ∇' induced on the hypersurface with respect to the normal is also the Riemannian connection of the hypersurface with the induced Riemannian metric.*

Proof. Since ∇ is the Riemannian connection of a Riemannian manifold with metric tensor g, ∇ has no torsion and

$$\nabla_X(g(Y, Z)) = g(\nabla_X Y, Z) + g(Y, \nabla_X Z)$$

for any vector fields X, Y, Z of M.

Applying the operator $\nabla_{BX'}$ to

$$g'(Y', Z') = g(BY', BZ'),$$

we obtain

$$\begin{aligned}\nabla'_{X'}(g'(Y', Z')) &= \nabla_{BX'}(g(BY', BZ')) \\ &= g(\nabla_{BX'}BY', BZ') + g(BY', \nabla_{BX'}BZ') \\ &= g(B\nabla'_{X'}Y' + h(X', Y')N, BZ') \\ &\quad + g(BY', B\nabla'_{X'}Z' + h(X', Z')N) \\ &= g(B\nabla'_{X'}Y', BZ') + g(BY', B\nabla'_{X'}Z') \\ &= g'(\nabla'_{X'}Y', Z') + g'(Y', \nabla'_{X'}Z'),\end{aligned}$$

that is,

$$\nabla'_{X'}(g'(Y', Z')) = g'(\nabla'_{X'}Y', Z') + g'(Y', \nabla'_{X'}Z'),$$

which, since ∇' has no torsion, shows that ∇' is the Riemannian connection of g'.

§5. Equations of Gauss and Weingarten

We know that

$$\nabla_{BX'}BY' = B\,\nabla'_{X'}Y' + h(X', Y')N. \tag{5.1}$$

Since $\nabla_{BX'}BY' - B\,\nabla'_{X'}Y'$ is linear in X' and Y', we put

$$(\nabla'B)(X', Y') = (\nabla'_{X'}B)Y' = \nabla_{BX'}BY' - B(\nabla'_{X'}Y'). \tag{5.2}$$

Then,

$$(\nabla'_{X'}B)Y' = h(X', Y')N \tag{5.3}$$

or

$$(\nabla'B)(X', Y') = h(X', Y')N.$$

In classical tensor calculus, $\nabla'B$ is called the *van der Waerden-Bortolotti covariant derivative* (Schouten [1]) of B along the hypersurface, and the above equation is called the *equation of Gauss*.

A hypersurface which satisfies

$$h = 0$$

is said to be *totally geodesic*, and a hypersurface which satisfies

$$h = \lambda g,$$

λ being a function on $i(P)$, is said to be *totally umbilical*.

PROPOSITION 5.1. *Let BX' be a vector field tangent to the hypersurface, and X and N extensions of BX' and N, respectively. Then $(\nabla_X N)\,|_{i(P)}$ does not depend on the extension. If we denote this vector field by $\nabla_{BX'}N$, then*

$$\nabla_{BX'}N = -BHX', \tag{5.4}$$

where H is a tensor field of type (1.1) *on the hypersurface such that*

$$h(X', Y') = g'(HX', Y'). \tag{5.5}$$

Proof. Denoting by $B_c{}^h X'^c(u)$ and $N^h(u)$ the components of BX' and N, and by $X^h(x)$ and $N^h(x)$ those of extensions of BX' and N, respectively, we have

$$X^h(x(u)) = B_c{}^h X'^c \quad \text{and} \quad N^h(x(u)) = N^h(u).$$

Thus

$$X^j\left(\partial_j N^h + \begin{Bmatrix} h \\ ji \end{Bmatrix} N^i\right)$$

are the components of $\nabla_X N$, so

$$\left.\left\{X^j\left(\partial_j N^h + \begin{Bmatrix} h \\ ji \end{Bmatrix} N^i\right)\right\}\right|_{x=x(u)} = \left.\left\{B_c{}^j X'^c\left(\partial_j N^h + \begin{Bmatrix} h \\ ji \end{Bmatrix} N^i\right)\right\}\right|_{x=x(u)}$$
$$= X'^c\left(\partial_c N^h + \begin{Bmatrix} h \\ ji \end{Bmatrix} B_c{}^j N^i\right),$$

from which we see that $\nabla_X N$ does not depend on the extension.

Since $\nabla_{BX'}N$ is a vector field defined along the hypersurface, it can be expressed in the form

$$\nabla_{BX'}N = BW' + \mu' N,$$

where W' is a vector field and μ' is a function on the hypersurface depending on X'. If we replace X' by $\alpha X'$, α being a function on the hypersurface, then

$$\nabla_{\alpha BX'}N = \alpha\, \nabla_{BX'}N = \alpha(BW' + \mu' N).$$

Thus, setting

$$\nabla_{\alpha BX'}N = B\overline{W}' + \bar{\mu}'N,$$

we have

$$\overline{W}' = \alpha W', \qquad \bar{\mu}' = \alpha\mu'.$$

These equations show that W' is of the form

$$W' = -HX',$$

where H is a tensor field of type (1.1) on the hypersurface and μ' has the form

$$\mu' = \pi(X'),$$

where π is a 1-form on the hypersurface.

Thus

$$\nabla_{BX'}N = -BHX' + \pi(X')N. \tag{5.6}$$

Applying the operator $\nabla_{BX'}$ to

$$g(BY', N) = 0,$$

we find

$$g(\nabla_{BX'}BY', N) + g(BY', \nabla_{BX'}N) = 0.$$

Substituting Eqs. (5.1) and (5.6) into this equation, we find

$$g(B\nabla'_{X'}Y' + h(X', Y')N, N) + g(BY', -BHX' + \pi(X')N) = 0,$$

from which

$$h(X', Y') - g(BY', BHX') = 0$$

or

$$h(X', Y') = g'(HX', Y').$$

Applying the operator $\nabla_{BX'}$ to

$$g(N, N) = 1,$$

we find

$$g(\nabla_{BX'}N, N) = 0.$$

Substituting Eq. (5.6) into this equation, we have

$$g(-BHX' + \pi(X')N, N) = 0,$$

from which

$$\pi(X') = 0;$$

thus we get Eq. (5.4). Equation (5.4) is called the *equation of Weingarten*.

The equation of Weingarten shows that the normal N of a totally geodesic hypersurface is parallel with respect to the connection ∇ of M.

§6. Equations of Gauss and Codazzi

The curvature tensor of the Riemannian manifold M is given by

$$K(X, Y)Z = \nabla_X\nabla_Y Z - \nabla_Y\nabla_X Z - \nabla_{[X,Y]}Z. \tag{6.1}$$

Setting

$$X = BX', \qquad Y = BY', \qquad Z = BZ',$$

we have

$$K(BX', BY')BZ' = \nabla_{BX'}\nabla_{BY'}BZ' - \nabla_{BY'}\nabla_{BX'}BZ' - \nabla_{B[X',BY']}BZ',$$

that is,

$$K(BX', BY')BZ' = \nabla_{BX'}\nabla_{BY'}BZ' - \nabla_{BY'}\nabla_{BX'}BZ' - \nabla_{[BX',Y']}BZ',$$

because of Proposition 3.1. Thus, applying the Gauss equation (5.1),

$$\begin{aligned} K(BX', BY')BZ' &= \nabla_{BX'}(B\nabla'_{Y'}Z' + h(Y', Z')N) \\ &\quad - \nabla_{BY'}(B\nabla'_{X'}Z' + h(X', Z')N) - (B\nabla'_{[X',Y']}Z' + h([X', Y'], Z')N). \end{aligned}$$

Again, by Eq. (5.1) and the equation of Weingarten,

$$\begin{aligned} &K(BX', BY')BZ' \\ &\quad = B\nabla'_{X'}\nabla'_{Y'}Z' + h(X', \nabla'_{Y'}Z')N + \nabla'_{X'}(h(Y', Z'))N - h(Y', Z')BHX' \\ &\qquad - B\nabla'_{Y'}\nabla'_{X'}Z' - h(Y', \nabla'_{X'}Z')N - \nabla'_{Y'}(h(X', Z'))N \\ &\qquad + h(X', Z')BHY' - (B\nabla'_{[X',Y']}Z' + h([X', Y'], Z')N) \\ &\quad = B(\nabla'_{X'}\nabla'_{Y'}Z' - \nabla'_{Y'}\nabla'_{X'}Z' - \nabla'_{[X',Y']}Z' \\ &\qquad - h(Y', Z')HX' + h(X', Z')HY') \\ &\qquad + ((\nabla'_{X'}h)(Y', Z') - (\nabla'_{Y'}h)(X', Z') \\ &\qquad + h(\nabla'_{X'}Y' - \nabla'_{Y'}X' - [X', Y'], Z'))N, \end{aligned}$$

that is,

$$\begin{aligned} K(BX', BY')BZ' &= B(K'(X', Y')Z' - h(Y', Z')HX' + h(X', Z')HY') \\ &\quad + ((\nabla'h)(X', Y', Z') - (\nabla'h)(Y', X', Z'))N, \end{aligned} \tag{6.2}$$

$K'(X', Y')Z'$ being the curvature tensor of the hypersurface. Putting

$$K(BX', BY', BZ', U) = g(K(BX', BY')BZ', U), \tag{6.3}$$

$$K'(X', Y', Z', U') = g'(K'(X', Y')Z', U'), \tag{6.4}$$

we find, from (6.2),

$$\begin{aligned} K(BX', BY', BZ', BU') = K'(X', Y', Z', U') - \{&h(X', U')h(Y', Z') \\ &- h(Y', U')h(X', Z')\} \end{aligned} \tag{6.5}$$

and

$$K(BX', BY', BZ', N) = (\nabla' h)(X', Y', Z') - (\nabla' h)(Y', X', Z'). \quad (6.6)$$

Equation (6.5) is called the *equation of Gauss* and Eq. (6.6) is called the *equation of Codazzi*.

If we put

$$X = BX', \qquad Y = BY', \qquad Z = N$$

in (6.1), we find

$$K(BX', BY')N = \nabla_{BX'}\nabla_{BY'}N - \nabla_{BY'}\nabla_{BX'}N - \nabla_{[BX', BY']}N,$$

that is,

$$K(BX', BY')N = \nabla_{BX'}\nabla_{BY'}N - \nabla_{BY'}\nabla_{BX'}N - \nabla_{B[X', Y']}N.$$

Thus, applying the equations of Gauss and Weingarten,

$$\begin{aligned} K(BX', BY')N \\ &= \nabla_{BX'}(-BHY') - \nabla_{BY'}(-BHX') - (-BH[X', Y']) \\ &= -B\nabla'_{X'}(HY') - h(X', HY')N \\ &\quad + B\nabla'_{Y'}(HX') + h(Y', HX')N + BH[X', Y'] \\ &= -B((\nabla'_{X'}H)Y' - (\nabla'_{Y'}H)X') - BH(\nabla'_{X'}Y' - \nabla'_{Y'}X' - [X', Y']) \\ &\quad - (h(X', HY') - h(Y', HX'))N, \end{aligned}$$

that is,

$$K(BX', BY')N = -B((\nabla'_{X'}H)Y' - (\nabla'_{Y'}H)X'), \quad (6.7)$$

by virtue of

$$\nabla'_{X'} Y' - \nabla'_{Y'} X' - [X', Y'] = 0$$

and

$$h(X', HY') = h(Y', HX').$$

It can readily be verified that Eq. (6.7) yields the equation of Codazzi also.

PROPOSITION 6.1. *A totally umbilical hypersurface in a space of constant curvature is of constant curvature.*

Proof. The curvature tensor of a space of constant curvature satisfies

$$K(X, Y, Z, U) = k(g(X, U)g(Y, Z) - g(Y, U)g(X, Z)),$$

k being a constant, and the second fundamental tensor $h(X', Y')$ of a totally umbilical hypersurface satisfies

$$h(X', Y') = \lambda g'(X', Y'),$$

λ being a function and consequently we have, from the Gauss equation (6.5),

$$k\{g(BX', BU')g(BY', BZ') - g(BY', BU')g(BX', BZ')\}$$
$$= K'(X', Y', Z', U') - \lambda\{g'(X', U')g'(Y', Z') - g'(Y', U')g'(X', Z')\}.$$

Hence,

$$K'(X', Y', Z', U') = (k + \lambda)\{g'(X', U')g'(Y', Z') - g'(Y, U')g'(X', Z')\},$$

which shows that the hypersurface has constant curvature $k + \lambda$ for $n - 1 > 2$. Consequently $\lambda = \text{const}$.

PROPOSITION 6.2. *For a hypersurface in a space of constant curvature, the tensor* $(\nabla' h)(X', Y', Z')$ *is symmetric in* X' *and* Y'.

Proof. The curvature tensor of a space of constant curvature satisfies

$$K(X, Y, Z, U) = k\{g(X, U)g(Y, Z) - g(Y, U)g(X, Z)\},$$

k being a constant, and consequently we have, from the Codazzi equation (6.6),

$$k\{g(BX', N)g(BY', BZ') - g(BY', N)g(BX', BZ')\}$$
$$= (\nabla' h)(X', Y', Z') - (\nabla' h)(Y', X', Z'),$$

that is,

$$(\nabla' h)(X', Y', Z') = (\nabla' h)(Y', X', Z'). \tag{6.8}$$

If we choose n orthonormal vectors

$$X_1 = N, \qquad X_\alpha = BX'_\alpha \quad (\alpha = 2, 3, \ldots, n),$$

then from the equation of Codazzi,

$$\sum_{\alpha=2}^{n} K(BX', BX'_\alpha, BX'_\alpha, N) = \sum_{\alpha=2}^{n} [(\nabla' h)(X', X'_\alpha, X'_\alpha) - (\nabla' h)(X'_\alpha, X', X'_\alpha)],$$

that is,

$$R(BX', N) = \nabla'_{X'} p_1 + (\delta' h)(X'), \tag{6.9}$$

where

$$p_1 = \langle h, g' \rangle. \tag{6.10}$$

If we denote by $k_1, \ldots, k_{n-1}$ the principal curvatures of P, that is, the roots of the characteristic equation

$$|h - kg| = 0, \tag{6.11}$$

then the first mean curvature M_1 and the second mean curvature M_2 of the hypersurface P are given by

$$(n-1)M_1 = \sum_{a=1}^{n-1} k_a = p_1 \tag{6.12}$$

and

$$\binom{n-1}{2} M_2 = \sum_{c<b}^{1\cdots n-1} k_c k_b = \tfrac{1}{2}(p_1^2 - p_2), \tag{6.13}$$

where

$$p_2 = \langle h, h \rangle. \tag{6.14}$$

Problems

1. Prove that the vectors

$$\nabla_j X_i = \partial_j X_i - \begin{Bmatrix} h \\ ji \end{Bmatrix} X_h$$

in E are all normal to the hypersurface V, and so, putting

$$(\nabla_j X_i) \cdot N = h_{ji},$$

we have the equations of Gauss,

$$\nabla_j X_i = h_{ji} N.$$

2. Prove that the vectors

$$\nabla_j N = \partial_j N$$

in E are all tangent to the hypersurface V and so we obtain the equations of Weingarten,

$$\nabla_j N = -h_j{}^h X_h.$$

3. Prove the Ricci identities

$$\nabla_k \nabla_j X_i - \nabla_j \nabla_k X_i = -K_{kji}{}^h X_h,$$

where

$$K_{kji}{}^h = \partial_k \begin{Bmatrix} h \\ ji \end{Bmatrix} - \partial_j \begin{Bmatrix} h \\ ki \end{Bmatrix} + \begin{Bmatrix} h \\ kt \end{Bmatrix}\begin{Bmatrix} t \\ ji \end{Bmatrix} - \begin{Bmatrix} h \\ jt \end{Bmatrix}\begin{Bmatrix} t \\ ki \end{Bmatrix},$$

and

$$\nabla_k \nabla_j N - \nabla_j \nabla_k N = 0.$$

4. Use the results of Problems 1–3 to obtain the equations of Gauss,

$$K_{kji}{}^t = h_k{}^t h_{ji} - h_j{}^t h_{ki},$$

and those of Codazzi,

$$\nabla_k h_{ji} - \nabla_j h_{ki} = 0.$$

5. Prove that if we put

$$z_i = X \cdot X_i, \qquad \alpha = X \cdot N,$$

then

$$\nabla_j z_i = g_{ji} + \alpha h_{ji},$$

$$\nabla_j \alpha = -h_{ji} z^i.$$

6. Prove that

$$M_1^2 - M_2 = \frac{1}{n^2(n-1)} \sum_{j<i} (k_j - k_i)^2.$$

7. When a compact and orientable hypersurface V of a Euclidean space E is totally umbilical, we have

$$h_{ji} = kg_{ji},$$

and consequently, from the equations of Gauss and Weingarten,

$$\nabla_j X_i = kg_{ji} N, \qquad \nabla_j N = -kX_j.$$

Prove that k is a constant. Since

$$M_1 = k,$$

this constant cannot be zero. (See, (2.9).)

Then,

$$\nabla_j \left(X + \frac{1}{k} N \right) = 0.$$

Prove that V is a sphere.

8. Consider a compact and orientable totally umbilical hypersurface V: $h_{ji} = kg_{ji}$. Prove that if we take a fixed vector C and put

$$f = C \cdot N,$$

then f satisfies

$$\nabla_j \nabla_i f = -k^2 f g_{ji}.$$

Use Obata's theorem to prove that V is a sphere.

9. Prove that

$$M_1 M_l - M_{l+1} = \frac{l!\,(n-l-1)!}{nn!} \sum_{\substack{t_1 < t_2 \\ i_3 < \cdots < t_{l+1}}} (k_{i_1} - k_{i_2})^2 k_{i_3} \cdots k_{i_{l+1}}.$$

10. It is known (Hardy et al. [1]) that

$$M_l^2 - M_{l-1} M_{l+1} \geq 0,$$

equality holding if and only if

$$k_1 = \cdots = k_n.$$

Prove that if $M_1, \cdots, M_l > 0$ then

$$M_1 M_l - M_{l-1} \geq 0,$$

equality holding if and only if

$$k_1 = \cdots = k_n.$$

11. Use the result of Problem 10 to generalize Theorem 2.2 to

THEOREM. *If a closed orientable hypersurface of an $(n + 1)$-dimensional Euclidean space satisfies $M_1, M_2, \cdots, M_l > 0$, $M_l = \text{const}$, and if the inner product α of the position vector X and the normal N has constant sign, then V is totally umbilical, and hence a sphere.*

12. Prove that Eq. (6.7) also gives the equation of Codazzi.

13. Suppose that a hypersurface of a space of constant curvature is totally umbilical:

$$h = \lambda g',$$

λ being a function on the hypersurface. Prove that λ is a constant.

14. Prove that

$$\nabla' p_1 + \delta' h = 0$$

for a hypersurface of a space of constant curvature.

CHAPTER

6

Closed Hypersurfaces of a Riemannian Manifold with Constant Mean Curvature

In this chapter, closed orientable hypersurfaces of a Riemannian manifold with constant mean curvature are studied.

For the generalization of the results stated in this chapter to general submanifolds of Riemannian manifolds, see Katsurada [4], Katsurada and Kôjyô [1], Katsurada and Nagai [1], Nagai [1], Okumura [1, 2], Yano and Okumura [1], Yano and Tani [2], and Yano [22].

§1. Hypersurfaces of a Riemannian Manifold Admitting an Infinitesimal Conformal Transformation

In this section let M be a Riemannian manifold admitting an infinitesimal conformal transformation X:

$$\pounds_X g = 2\rho g. \tag{1.1}$$

If $\rho = c =$ const, then the infinitesimal transformation X is homothetic, and if $\rho = 0$, then

$$\pounds_X g = 0, \tag{1.2}$$

so the transformation X is an isometry.

If $i(P)$ is a hypersurface in M, then, along the hypersurface $i(P)$,

$$X = BX' + \alpha N, \tag{1.3}$$

X' being a vector field in P and α a function on P.

Since

$$\nabla_{BY'}X = h(Y', X')N + B\nabla'_{Y'}X' + (\nabla'_{X'}\alpha)N - \alpha BHY',$$

we have

$$(\pounds_X g)(BY', BZ') = (\pounds_{X'}g')(Y', Z') - 2\alpha h(Y', Z'). \tag{1.4}$$

Substituting

$$BX' = X - \alpha N$$

into Eq. (6.9) of Chapter 5, we find

$$R(X - \alpha N, N) = \nabla'_{X'}p_1 + \delta(HX') + \langle h, \nabla' X'\rangle,$$

and consequently

$$R(X, N) - \alpha R(N, N) = \nabla'_{X'}p_1 + \delta(HX') + \tfrac{1}{2}\langle h, \pounds_{X'}g'\rangle \tag{1.5}$$

or

$$R(X, N) - \alpha R(N, N) = \nabla'_{X'}p_1 + \delta'(HX') + \alpha p_2 + \tfrac{1}{2}\langle h, L\rangle, \tag{1.6}$$

where

$$p_2 = |h|^2 \tag{1.7}$$

and

$$L(Y', Z') = (\pounds_X g)(BY', BZ'). \tag{1.8}$$

If X is an infinitesimal conformal transformation, then from Eq. (1.6),

$$R(X, N) - \alpha R(N, N) = \nabla'_{X'}p_1 + \delta(HX') + \alpha p_2 + \rho p_1. \tag{1.9}$$

If X is an infinitesimal homothetic transformation, then

$$R(X, N) - \alpha R(N, N) = \nabla'_{X'}p_1 + \delta'(HX') + \alpha p_2 + c p_1, \tag{1.10}$$

and if X is an infinitesimal isometry, then

$$R(X, N) - \alpha R(N, N) = \nabla'_{X'}p_1 + \delta'(HX') + \alpha p_2. \tag{1.11}$$

§2. Integral Formulas

Define 1-forms ω and ω' by

$$\omega(Z) = g(X, Z) \tag{2.1}$$

and

$$\omega'(Z') = g'(X', Z'), \tag{2.2}$$

respectively, for any vector fields Z and Z'. Then

$$\begin{aligned}\omega'(Z') = g'(X', Z') &= g(BX', BZ')\\ &= g(X - \alpha N, BZ')\\ &= g(X, BZ'),\end{aligned}$$

that is,

$$\omega'(Z') = \omega(BZ'). \tag{2.3}$$

From this, by covariant differentiation along the hypersurface,

$$(\nabla'\omega')(Y', Z') = (\nabla\omega)(BY', BZ') + \omega(h(Y', Z')N),$$

or

$$-\delta\omega' = \tfrac{1}{2}\langle L, g'\rangle + \alpha p_1. \tag{2.4}$$

Thus, assuming P to be compact and orientable and integrating over P,

$$\frac{1}{2}\int_P \langle L, g'\rangle\, dS + (n-1)\int_P \alpha M_1\, dS = 0, \tag{2.5}$$

dS being the surface element of P.

If the vector field X is conformal, that is, if $\pounds_X g = 2\rho g$, then, from the formula above,

$$\int_P \rho\, dS + \int_P \alpha M_1\, dS = 0. \tag{2.6}$$

We next consider the 1-form

$$\theta' = \omega \circ H.$$

We have, from this, by covariant differentiation along the hypersurface,

$$\nabla'\theta' = \nabla'(\omega \circ H),$$

from which

$$\delta'\theta' = \delta'(HX'). \tag{2.7}$$

Thus, from (1.6) and (2.7), we find

$$\delta'\theta' = R(X, N) - \alpha R(N, N) - \tfrac{1}{2}\langle L, h\rangle - \nabla'_{X'} p_1 - \alpha p_2.$$

But,

$$p_1 = (n-1)M_1, \qquad p_2 = (n-1)^2 M_1 - (n-1)(n-2)M_2,$$

and consequently

$$\begin{aligned}\delta'\theta' = R(X, N) &- \alpha R(N, N) - \tfrac{1}{2}\langle L, h\rangle\\ &- (n-1)\nabla'_{X'} M_1 - (n-1)\alpha\{(n-1)M_1^2 - (n-2)M_2\}.\end{aligned}$$

Thus, assuming P to be compact and orientable,

$$\int_P [\tfrac{1}{2}\langle L, h\rangle - R(X, N) + \alpha R(N, N) + (n-1)\nabla'_{X'}M_1 + (n-1)\alpha \times \{(n-1)M_1^2 - (n-2)M_2\}]\, dS = 0. \quad (2.8)$$

If the vector field X is conformal, then, from the equation above,

$$\int_P [R(X, N) - \alpha R(N, N)]\, dS = (n-1)\int_P [\nabla'_{X'}M_1 + \rho M_1 + \alpha\{(n-1)M_1^2 - (n-2)M_2\}]\, dS. \quad (2.9)$$

Covariant differentiation of the function

$$\alpha = \langle X, N\rangle \quad (2.10)$$

along the hypersurface yields

$$\begin{aligned}(\nabla'\alpha)(Z') &= \langle(\nabla X)BZ', N\rangle + \langle X, \nabla_{BZ'}N\rangle \\ &= \langle(\nabla X)BZ', N\rangle - \langle X, BHZ'\rangle,\end{aligned}$$

or

$$(\nabla'\alpha)(Z') = \langle(\nabla X)BZ', N\rangle - \langle HX', Z'\rangle.$$

Taking the second covariant derivative of α along P, we get

$$(\nabla'\nabla'\alpha)(Y', Z') = \langle(\nabla\nabla X)(BY', BZ'), N\rangle + \langle(\nabla X)N, N\rangle h(Y', Z') - \langle(\nabla X)BZ', BHY'\rangle - (\nabla'_{Y'}(HX'))Z',$$

from which

$$\begin{aligned}-\delta'\alpha = {}&\langle Q', g'N\rangle - R(X, N) + \tfrac{1}{2}p_1(\pounds_X g)(N, N) \\ &-\tfrac{1}{2}\langle L, h\rangle + \delta'(HX'),\end{aligned} \quad (2.11)$$

by virtue of

$$(\nabla\nabla X)(Y, Z) = (\pounds_X \nabla)(Y, Z) - K(X, Y)Z,$$

where

$$Q'(Y', Z') = (\pounds_X \nabla)(BY', BZ'). \quad (2.12)$$

Thus, assuming P to be compact and orientable, we get, from (2.11),

$$\int_P [\langle Q', g'N\rangle - R(X, N) + \tfrac{1}{2}p_1(\pounds_X g)(N, N) - \tfrac{1}{2}\langle L, h\rangle]\, dS = 0. \quad (2.13)$$

If the vector field X is a conformal vector field,

$$\pounds_X g = 2\rho g$$

and

$$(\pounds_X \nabla)(Y, Z) = (d\rho)(Y)Z + (d\rho)(Z)Y - g(Y, Z)D\rho,$$

from which

$$
\begin{aligned}
(\pounds_X g)(N, N) &= 2\rho, \\
L(Y', Z') = (\pounds_X g)(BY', BZ') &= 2\rho g'(Y', Z'), \\
\langle L, h\rangle &= 2(n-1)\rho H_1, \\
Q'(Y', Z') &= (\pounds_X \nabla)(BY', BZ') \\
&= (d\rho)(BY')BZ' + (d\rho)(BZ')BY' - g'(Y', Z')D\rho, \\
\langle Q', g'N\rangle &= -(n-1)(d\rho)(N).
\end{aligned}
$$

Thus, from (2.13),

$$\int_P [(n-1)(d\rho)(N) + R(X, N)]\, dS = 0. \tag{2.14}$$

§3. Integral Formulas for the Case $M_1 = \text{const}$

We assume in this section that the Riemannian manifold M admits an infinitesimal conformal transformation X, the hypersurface is compact and orientable, and the first mean curvature M_1 of the hypersurface is constant. Then Eqs. (2.6) and (2.9) become, respectively,

$$\int_P \rho\, dS + M_1 \int_P \alpha\, dS = 0 \tag{3.1}$$

and

$$\frac{1}{n-1}\int_P [R(X, N) - \alpha R(N, N)]\, dS = M_1 \int_P \rho\, dS + \int_P \alpha\{(n-1)M_1^2 - (n-2)M_2\}\, dS. \tag{3.2}$$

Eliminating $\int_P \rho\, dS$ from these equations,

$$\frac{1}{n-1}\int_P [R(X, N) - \alpha R(N, N)]\, dS = (n-2)\int_P \alpha(M_1^2 - M_2)\, dS. \tag{3.3}$$

If the Riemannian manifold M is an Einstein space, then

$$R = \lambda g,$$

and consequently we have, from (3.3),

$$\int_P \alpha(M_1^2 - M_2)\, dS = 0, \tag{3.4}$$

where

$$M_1^2 - M_2^2 = \frac{1}{(n-1)^2(n-2)} \sum_{a \neq b} (k_a - k_b)^2. \tag{3.5}$$

Thus, if α has constant sign,

$$M_1^2 - M_2^2 = 0$$

so

$$k_1 = k_2 = \cdots = k_{n-1}.$$

Thus

THEOREM 3.1. *Let M be an n-dimensional orientable Einstein space and P a closed orientable hypersurface in M whose first mean curvature is constant. We suppose that M admits a 1-parameter group of conformal transformations such that the inner product α of the generating vector X and the normal N to the hypersurface has constant sign on P. Then every point of P is umbilical* (Katsurada [2]).

§4. Hypersurfaces With Constant First Mean Curvature of a Riemannian Manifold Admitting an Infinitesimal Homothetic Transformation

We assume in this section that the Riemannian manifold admits an infinitesimal homothetic transformation X and the first mean curvature of the hypersurface is constant. Then Eqs. (2.6), (2.9), and (2.14) become

$$c\int_P dS + M_1\int_P \alpha\, dS = 0, \tag{4.1}$$

$$\frac{1}{n-1}\int_P \alpha R(N, N)\, dS + cM_1\int_P dS + \int_P \alpha\{(n-1)M_1^2 - (n-2)M_2\}\, dS = 0 \tag{4.2}$$

and

$$\int_P R(X, N)\, dS = 0, \tag{4.3}$$

respectively.

Eliminating $\int_P dS$ from (4.1) and (4.2),

$$\int_P \alpha\left[\frac{1}{n-1} R(N, N) + (n-2)(M_1^2 - M_2)\right] dS = 0. \tag{4.4}$$

Thus, if α has constant sign and $R(N, N) \geq 0$, then

$$M_1^2 - M_2 = 0,$$

so

$$k_1 = k_2 = \cdots = k_{n-1}.$$

Thus,

THEOREM 4.1. *Let M be an n-dimensional orientable Riemannian manifold and P a closed orientable hypersurface in M whose first mean curvature is constant. We suppose that M admits a 1-parameter group of homothetic transformations such that the inner product of the generating vector X and the normal N to the hypersurface has constant sign on P, and the Ricci curvature $R(N, N)$ with respect to the normal N is nonnegative on P. Then every point of P is umbilical and $R(N, N) = 0$ on P* (Yano [16]).

Assume now that the Riemannian manifold M under consideration is an Einstein space: $R = \lambda g$, λ being a constant. Then, from (4.3),

$$\lambda \int_P \alpha \, dS = 0.$$

Thus, if α has constant sign on P, we have $\lambda = 0$, and hence $R = 0$. Thus,

COROLLARY 4.2. *Let M be an n-dimensional orientable Einstein space and P a closed orientable hypersurface in M whose first mean curvature is constant. Then, if M admits a 1-parameter group of homothetic transformations such that the inner product of the generating vector X and the normal N to P has constant sign on P, the scalar curvature of the space vanishes and every point of the hypersurface is umbilical* (Yano [16]).

If $\alpha = 0$, then (2.6) becomes

$$c \int_P dS = 0,$$

from which

$$c = 0.$$

Thus,

COROLLARY 4.3. *Let M be an n-dimensional orientable Riemannian manifold and P a closed orientable hypersurface in M. If M admits a 1-parameter group of homothetic transformations such that the generating vector field X is always tangent to P, then the 1-parameter group is a group of isometries* (Yano [16]).

§5. Integral Formulas in M Admitting a Function v Such That $\nabla\nabla v = f(v)g$

We assume in this section that the Riemannian manifold M admits a nonconstant function v such that

$$\nabla\nabla v = f(v)g, \tag{5.1}$$

where $f(v)$ is a real-valued differentiable function of v.

We denote by X the vector field associated with the 1-form $\xi = \nabla v$, that is,

$$g(X, Y) = \xi(Y) \tag{5.2}$$

for any vector field Y. We put

$$X = BX' + \alpha N. \tag{5.3}$$

Then, from (5.1),

$$(\nabla\nabla v)(BY', BZ') = f(v)g(BY', BZ'),$$

$$\nabla_{BY'}(\nabla_{BZ'}v) - (\nabla v)(\nabla_{BY'}BZ') = f(v)g'(Y', Z'),$$

$$\nabla_{BY'}((\nabla v)(Z')) - (\nabla v)(B\nabla'_{Y'}Z' + h(Y', Z')N) = f(v)g'(Y', Z'),$$

that is,

$$(\nabla'\nabla' v)(Y', Z') = f(v)g'(Y', Z') + \alpha h(Y', Z'), \tag{5.4}$$

from which

$$\Delta' v = -(n-1)f(v) - \alpha p_1. \tag{5.5}$$

Also, from (5.1),

$$\begin{aligned} (\nabla\nabla v)(BY', N) &= f(v)g(BY', N), \\ \nabla_{BY'}((\nabla v)(N)) - (\nabla v)(\nabla_{BY'}N) &= 0, \\ \nabla'_{Y'}\alpha + \xi(BHY') &= 0, \\ \nabla'_{Y'}\alpha &= -\xi'(HY'). \end{aligned} \tag{5.6}$$

On the other hand, Eq. (5.1) may be written as

$$\nabla_Z X = f(v)Z,$$

from which

$$\nabla_Y\nabla_Z X = f'(v)\xi(Y)Z + f(v)\nabla_Y Z.$$

Consequently,

$$\nabla_Y\nabla_Z X - \nabla_Z\nabla_Y X - \nabla_{[Y,Z]}X = f'(v)(\xi(Y)Z - \xi(Z)Y) + f(v)(\nabla_Y Z - \nabla_Z Y) - f(v)[Y, Z],$$

that is,

$$K(Y, Z)X = f'(v)(\xi(Y)Z - \xi(Z)Y).$$

Thus,

$$K(Y, Z, X, U) = g(K(Y, Z)X, U)$$
$$= f'(v)[g(Y, X)g(Z, U) - g(Z, X)g(Y, U)],$$

from which

$$R(Z, X) = -(n-1)f'(v)g(Z, X).$$

Consequently,

$$R(X, N) = -(n-1)f'(v)\alpha,$$

which may also be written as

$$R(BX' + \alpha N, N) = -(n-1)f'(v)\alpha,$$

or, by virtue of formula (6.9) of Chapter 5,

$$\nabla'_{X'}p_1 + (\delta'h)(X') + \alpha R(N, N) = -(n-1)f'(v)\alpha,$$

that is,

$$\alpha R(N, N) + (n-1)f'(v)\alpha + \nabla'_{X'}p_1 + \delta(HX) + f(v)p_1 + \alpha p_2 = 0, \quad (5.7)$$

by virtue of

$$\langle h, \nabla'\nabla'v'\rangle = f(v)p_1 + \alpha p_2,$$

obtained from (5.4).

Assuming that the hypersurface P is closed and orientable, integrate (5.5) and (5.7) over P. Then

$$(n-1)\int_P f(v)\, dS + \int_P \alpha p_1\, dS = 0, \quad (5.8)$$

and

$$\int_P [\alpha R(N, N) + (n-1)f'(v)\alpha + \nabla'_{X'}p_1 + f(v)p_1 + \alpha p_2]\, dS = 0. \quad (5.9)$$

Moreover, assume that the first mean curvature of P is constant:

$$M_1 = \frac{1}{n-1}\, p_1 = \text{const.}$$

Thus, from (5.8) and (5.9),

$$(n-1)\int_P f(v)\, dS + p_1\int_P \alpha\, dS = 0$$

and

$$\int_P \alpha R(N, N)\, dS + (n-1)\int_P f'(v)\alpha\, dS + p_1\int_P f(v)\, dS + \int_P \alpha p_2\, dS = 0,$$

respectively. Eliminating $\int_P f(v)\,dS$ from these two equations, we find

$$\int_P \alpha\left[(R+(n-1)f'g)(N,N)+\left(p_2-\frac{1}{n-1}\,p_1^2\right)\right]dS=0,$$

$$\int_P \alpha\left[(R+(n-1)f'g)(N,N)+\left|h-\frac{1}{n-1}\,p_1 g\right|^2\right]dS=0, \quad (5.10)$$

by virtue of

$$\begin{aligned}\left|h-\frac{1}{n-1}\,p_1 g'\right|^2 &= \left\langle h-\frac{1}{n-1}\,p_1 g',\, h-\frac{1}{n-1}\,p_1 g'\right\rangle \\ &= p_2-\frac{2}{n-1}\,p_1^2+\frac{1}{n-1}\,p_1^2 \\ &= p_2-\frac{1}{n-1}\,p_1^2.\end{aligned}$$

§6. Hypersurfaces with Constant First Mean Curvature

In this section we first prove

THEOREM 6.1. *Let M be an orientable Riemannian manifold of dimension n admitting a nonconstant function v such that*

$$\nabla\nabla v = f(v)g, \quad (6.1)$$

where f is a differentiable function of v and P is a closed and orientable hypersurface in M. If

(*i*) *the first mean curvature of P in M is constant,*

(*ii*) $[R+(n-1)f'(v)g](N,N)\geq 0$ *on P, where N is the unit normal to P, and*

(*iii*) *the inner product $\alpha=\langle\nabla v, N\rangle$ has constant sign on P,*

then every point of P is umbilical (Yano [18]).

Proof. If the three conditions of the theorem are satisfied,

$$[R+(n-1)f'(v)g](N,N)\geq 0,$$

$$\left|h-\frac{1}{n-1}\,p_1 g'\right|^2\geq 0$$

from formula (5.10) and α has constant sign on P; hence

$$h = \frac{1}{n-1} p_1 g',$$

showing that every point of P is umbilical.

THEOREM 6.2. *Let M be an orientable Riemannian manifold of dimension n admitting a nonconstant function v such that*

$$\nabla\nabla v = kvg, \quad k = \text{const}, \tag{6.2}$$

and let P be a closed orientable hypersurface in M with the properties,

(*i*) *its first mean curvature is constant,*

(*ii*) $[R + (n-1)kg](N, N) \geq 0$ *on P, and*

(*iii*) *the inner product $\langle \nabla v, N \rangle$ has constant sign on P.*

Then every point of P is umbilical. If, moreover, v is not constant on P, then P is isometric to a sphere (Yano [18]).

Proof. The first part of Theorem 6.2 is a special case of Theorem 6.1 with $f(v) = kv$, k being a constant.

If

$$v \neq \text{const} \tag{6.3}$$

along P, then, since P is umbilical, that is,

$$h = \lambda g', \quad \lambda = \text{const}, \tag{6.4}$$

$$\nabla'\nabla' v = (kv + \lambda\alpha)g' \tag{6.5}$$

from (5.4) with $f(v) = kv$, and

$$\nabla'\alpha = -\lambda\xi = -\lambda\nabla' v \tag{6.6}$$

from (5.6). Hence

$$\alpha + \lambda v = c = \text{const}.$$

Substituting this into (6.5), we find

$$\nabla'\nabla' v = [kv + \lambda(c - \lambda v)]g'$$

or

$$\nabla'\nabla' v = [-(\lambda^2 - k)v + \lambda c]g'. \tag{6.7}$$

Here $\lambda^2 - k \neq 0$. For if $\lambda^2 - k = 0$ then (6.7) becomes $\nabla'\nabla' v = \lambda c g'$, from which $\Delta' v = -(n-1)\lambda c$, which is impossible unless $v = \text{const}$.

Thus, $\lambda^2 - k$ being different from zero, then from (6.7)

$$\nabla'\nabla'\left(v - \frac{\lambda c}{\lambda^2 - k}\right) = -(\lambda^2 - k)\left(v - \frac{\lambda c}{\lambda^2 - k}\right)g', \tag{6.8}$$

from which

$$\Delta'\left(v - \frac{\lambda c}{\lambda^2 - k}\right) = (n-1)(\lambda^2 - k)\left(v - \frac{\lambda c}{\lambda^2 - k}\right). \tag{6.9}$$

Consequently,

$$\lambda^2 - k \geq 0.$$

By a theorem of Obata (see Theorem 6.2 in Chapter 1), the hypersurface P is isometric to a sphere.

THEOREM 6.3. *Let M be an orientable Riemannian manifold of dimension n admitting nonconstant function v such that*

$$\nabla\nabla v = kg, \quad k = \text{const}, \tag{6.10}$$

and let P be a closed hypersurface with the properties

(*i*) *its first mean curvature is constant,*

(*ii*) $R(N, N) \geq 0$ *on P, and*

(*iii*) *the inner product $\langle \nabla v, N \rangle$ has constant sign on P.*

Then every point of P is umbilical. If, moreover, $v \neq$ const on P, then P is isometric to a sphere (Yano [18]).

Proof. The first part of Theorem 6.3 is a special case of Theorem 4.1 with $f(v) = k =$ const.

If

$$v \neq \text{const}$$

along P, then since P is umbilical, that is,

$$\begin{aligned} h &= \lambda g', \qquad \lambda = \text{const}, \\ \nabla'\nabla' v &= (k + \lambda\alpha)g', \end{aligned} \tag{6.11}$$

from (5.4) with $f(v) = k$, and

$$\nabla'\alpha = -\lambda\nabla' v, \tag{6.12}$$

from (5.6). Hence

$$\alpha + \lambda v = c = \text{const}. \tag{6.13}$$

Substituting (6.13) into (6.11), we find

$$\nabla'\nabla' v = (-\lambda^2 v + k + \lambda c)g'. \tag{6.14}$$

Here $\lambda \neq 0$. Because if $\lambda = 0$, then (6.11) becomes $\nabla'\nabla' v = kg'$, from which $\Delta' v = -(n-1)k$, which is impossible unless $v =$ const.

Thus, λ being different from zero,

$$\nabla'\nabla'\left(v - \frac{k + \lambda c}{\lambda^2}\right) = -\lambda^2\left(v - \frac{k + \lambda c}{\lambda^2}\right)g' \tag{6.15}$$

by (6.14), showing that P is isometric to a sphere.

Problems

1. Let X be a conformal Killing vector field and put

$$X = BX' + \alpha N.$$

Prove that

$$-\delta'\alpha = \langle Q', g'N\rangle - R(X, N) + \tfrac{1}{2}p_1(\pounds_X g)(N, N) - \tfrac{1}{2}\langle L, h\rangle + \delta'(HX').$$

(See (2.11).)

2. In Theorem 3.1 we assumed that an Einstein manifold admits an infinitesimal conformal transformation. Prove that in this case there exists a function ρ such that

$$\nabla\nabla\rho = f\cdot g.$$

3. Prove Eq. (5.5):

$$\Delta' v = -(n-1)f(v) - \alpha p_1.$$

4. Prove Eq. (5.7):

$$\alpha R(N, N) + (n-1)f'(v)\alpha + \nabla'_{X'}p_1 + \delta(HX) + f(v)p_1 + \alpha p_2 = 0.$$

CHAPTER

7

Harmonic 1-Forms and Killing Vector Fields in Riemannian Manifolds With Boundary

In this chapter we consider a compact and orientable Riemannian manifold M with compact orientable boundary P. For an arbitrary vector field X in M, we have *Stokes' theorem*:

$$\int_M \delta X \, dV = -\int_P \langle X, N \rangle \, dS,$$

where N is the unit normal to the boundary, dV is the volume element of M, and dS is the surface element of P.

§1. Harmonic 1-Forms in Riemannian Manifolds with Boundary

A vector field X, or its associated 1-form ξ, is said to be harmonic when it satisfies

$$d\xi = 0, \qquad \delta\xi = 0. \tag{1.1}$$

For a harmonic vector field X, we then have

$$\Delta X = 0. \tag{1.2}$$

By a straightforward computation, we have

$$\langle \Delta X, X \rangle - \tfrac{1}{2} |d\xi|^2 - (\delta\xi)^2 = -\delta[i(X)\, d\xi + (\delta\xi) \cdot \xi]. \tag{1.3}$$

Integrating both members of Eq. (1.3) over the manifold M and applying Stokes' theorem to the right-hand member, we get

$$\int_M [\langle \Delta X, X\rangle - \tfrac{1}{2}\,|d\xi|^2 - (\delta\xi)^2]\, dV = \int_P [(d\xi)(X, N) + (\delta\xi)\langle X, N\rangle]\, dS. \tag{1.4}$$

Suppose that X is a harmonic vector field. Then it satisfies

$$\Delta X = 0 \quad \text{in } M$$

and

$$(d\xi)(X, N) + (\delta\xi)\langle X, N\rangle = 0 \quad \text{on } P.$$

Conversely, if a vector field X satisfies these conditions, we have, from Eq. (1.4),

$$\int_M [\tfrac{1}{2}\,|d\xi|^2 + (\delta\xi)^2]\, dV = 0,$$

from which we obtain

$$d\xi = 0, \qquad \delta\xi = 0 \quad \text{in } M.$$

Thus we have

THEOREM 1.1. *A necessary and sufficient condition for a vector field X in M with boundary P to be harmonic is that*

$$\begin{cases} \Delta X = 0 & \textit{in } M, \\ (d\xi)(X, N) + (\delta\xi)\langle X, N\rangle = 0 & \textit{on } P \end{cases} \tag{1.5}$$

(Hsiung [3–6], Yano and Ako [2]).

If the vector field X vanishes on the boundary P, then the second condition of (1.5) is automatically satisfied. Thus we have

PROPOSITION 1.2. *A necessary and sufficient condition for a vector field X in M vanishing on the boundary P to be harmonic is that*

$$\Delta X = 0$$

in M.

Now, if, on the boundary P,

$$X = BX' + \alpha N, \tag{1.6}$$

then

$$\xi(BY') = \xi'(Y'), \qquad \xi(N) = \alpha \tag{1.7}$$

for any Y' of the boundary P, ξ' being a 1-form of P associated with X'.

Differentiating the first equation of (1.7) covariantly along the boundary and taking account of (1.7), we find

$$(\nabla\xi)(BX', BY') + \xi((\nabla_{BX'}B)Y') = (\nabla'\xi')(X', Y'),$$

that is,

$$(\nabla\xi)(BX', BY') + \alpha h(X', Y') = (\nabla'\xi')(X', Y'),$$

from which, by contraction,

$$\delta\xi + (\nabla\xi)(N, N) - \alpha p_1 = \delta'\xi' \tag{1.8}$$

by virtue of the relation

$$\sum_{a=1}^{n-1}(BX'_a) \otimes (BX'_a) = G - N \otimes N$$

for an orthonormal basis $\{X'_1, \ldots, X'_{n-1}\}$, G being the contravariant form of the metric tensor.

Differentiating the second equation of (1.7) covariantly along the boundary and taking account of (1.7), we find

$$(\nabla\xi)(BX', N) + \xi(\nabla_{BX'}N) = \nabla'_{X'}\alpha,$$

that is,

$$(\nabla\xi)(BX', N) - h(X', X') = \nabla'_{X'}\alpha$$

or

$$(\nabla\xi)(X, N) - \alpha(\nabla\xi)(N, N) - h(X', X') = \nabla'_{X'}\alpha. \tag{1.9}$$

From (1.8) and (1.9) we have, after eliminating $(\nabla\xi)(N, N)$,

$$(\nabla\xi)(X, N) = h(X', X') + \alpha^2 p_1 - \alpha\,\delta\xi + 2\alpha\,\delta'\xi' - \delta'(\alpha\xi'), \tag{1.10}$$

from which

$$\begin{aligned}(d\xi)(X, N) + \alpha(\delta\xi) = &-(\nabla\xi)(N, X) + h(X', X')\\ &+ \alpha^2 p_1 + 2\alpha\,\delta'\xi' - \delta'(\alpha\xi').\end{aligned} \tag{1.11}$$

Thus from Theorem 1.1 we obtain

PROPOSITION 1.3. *A necessary and sufficient condition for a vector field X in M with boundary P to be harmonic is that*

$$\begin{cases}\Delta X = 0 \quad \textit{in } M,\\ (\nabla\xi)(N, X) = h(X', X') + \alpha^2 p_1 + 2\alpha\,\delta'\xi' - \delta'(\alpha\xi') \quad \textit{on } P\end{cases} \tag{1.12}$$

(Yano and Ako [2]).

If the vector field X is tangential to the boundary P, then $\alpha = 0$, and consequently from Proposition 1.3 we obtain

PROPOSITION 1.4. *A necessary and sufficient condition for a vector field X in M tangential to the boundary P to be harmonic is that*

$$\begin{cases} \Delta X = 0 \quad \text{in } M, \\ (\nabla\xi)(N, X) = h(X', X') \quad \text{on } P \end{cases} \tag{1.13}$$

(Yano [13], Yano and Ako [2]).

If the vector field X is normal to the boundary P, then $X' = 0$ and $X = \alpha N$, and consequently

$$(d\xi)(X, N) + \alpha(\delta\xi) = (d\xi)(\alpha N, N) + \alpha(\delta\xi) = \alpha(\delta\xi).$$

Thus from Theorem 1.1 we obtain

PROPOSITION 1.5. *A necessary and sufficient condition for a vector field X in M normal to the boundary P to be harmonic is that*

$$\begin{cases} \Delta X = 0 \quad \text{in } M, \\ \alpha(\delta\xi) = 0 \quad \text{on } P \end{cases} \tag{1.14}$$

(Yano [13], Yano and Ako [2]).

Now, in general

$$\tfrac{1}{2}\Delta(g(X, X)) = g(\delta\nabla X, X) - |\nabla X|^2,$$

and

$$\tfrac{1}{2}\Delta(g(X, X)) = \tfrac{1}{2}\delta\nabla(g(X, X)).$$

Consequently,

$$g(\delta\nabla X, X) - |\nabla X|^2 = \tfrac{1}{2}\delta\nabla(g(X, X)),$$

from which, by integrating over M,

$$\int_M [g(\delta\nabla X, X) - |\nabla X|^2]\, dV = -\int_P (\nabla\xi)(N, X)\, dS. \tag{1.15}$$

Equations (1.11) and (1.15) give

$$\int_M [g(\delta\nabla X, X) - |\nabla X|^2]\, dV = \int_P [(d\xi)(X, N) + \alpha(\delta\xi)]\, dS$$
$$- \int_P [h(X', X') + \alpha^2 p_1 + 2\alpha(\delta'\xi')]\, dS. \tag{1.16}$$

Forming (1.4) − (1.16), we obtain

$$\int_M [R(X, X) + |\nabla X|^2 - \tfrac{1}{2}|d\xi|^2 - |\delta\xi|^2]\, dV$$
$$= \int_P [h(X', X') + \alpha^2 p_1 + 2\alpha(\delta'\xi')]\, dS. \tag{1.17}$$

Thus, for a harmonic vector X,

$$\int_M [R(X, X) + |\nabla X|^2]\, dV = \int_P [h(X', X') + \alpha^2 p_1 + 2\alpha(\delta'\xi')]\, dS,$$

from which we obtain

PROPOSITION 1.6. *If $R(X, X) \geq 0$, and if a harmonic vector field X satisfies one of the following alternate sets of conditions on P,*

(*i*) $h(X', X') \leq 0, \quad \alpha = 0,$

(*ii*) $h(X', X') \leq 0, \quad p_1 \leq 0, \quad \delta'\xi' = 0,$

(*iii*) $X' = 0, \quad p_1 \leq 0,$

then

$$R(X, X) = 0, \qquad \nabla X = 0 \quad \text{in } M,$$

and

$$h(X', X') = 0, \qquad p_1 = 0 \quad \text{on } P.$$

Thus, if $R(X, X)$ is positive-definite, there is no such vector field other than the zero vector (Bochner [13]).

PROPOSITION 1.7. *If $R(X, X)$ is positive-definite and $h(X', X') \leq 0$, then there is no harmonic vector tangential to the boundary other than the zero vector.*

PROPOSITION 1.8. *If $R(X, X)$ is positive-definite and $p_1 \leq 0$, then there is no harmonic vector normal to the boundary other than the zero vector.*

Duff and Spencer [1] showed that the absolute first Betti number $b_1(M)$ is equal to the number of linearly independent (with real coefficients) harmonic 1-forms tangential to the boundary P, and the relative first Betti number of M modulo P, $b_1(M, P)$, is equal to the number of linearly independent (with real coefficients) harmonic 1-forms normal to the boundary P. Thus

PROPOSITION 1.9. *If $R(X, X)$ is positive-definite and $h(X', X') \leq 0$, then*

$$b_1(M) = b_{n-1}(M, P) = 0.$$

PROPOSITION 1.10. *If $R(X, X)$ is positive-definite and $p_1 \leq 0$, then*

$$b_1(M, P) = b_{n-1}(M) = 0$$

(Nakae [1]).

§2. Killing Vector Fields in Riemannian Manifolds with Boundary

A vector field X is said to be a Killing vector field if it satisfies

$$\pounds_X g = 0, \tag{2.1}$$

and consequently

$$\delta\xi = 0. \tag{2.2}$$

From (2.1),

$$\pounds_X \nabla = 0,$$

from which

$$\square X = 0. \tag{2.3}$$

(For the definition of $\square X$, see Eq. (1.7) of Chapter 2.)

By a straightforward computation,

$$\langle \square X, X\rangle - \tfrac{1}{2}|\pounds_X g|^2 + (\delta\xi)^2 = \delta[i(X)(\pounds_X g) + (\delta\xi)\cdot\xi]. \tag{2.4}$$

Integrating both members of Eq. (2.4) over the manifold M and applying Stokes' theorem to the right-hand member, we get

$$\int_M [\langle \square X, X\rangle - \tfrac{1}{2}|\pounds_X g|^2 + (\delta\xi)^2]\, dV$$
$$= -\int_P [(\pounds_X g)(X, N) + (\delta\xi)\xi(N)]\, dS. \tag{2.5}$$

Suppose that X is a Killing vector field. Then it satisfies

$$\square X = 0, \qquad \delta\xi = 0 \quad \text{in } M$$

and

$$(\pounds_X g)(X, N) = 0 \quad \text{on } P.$$

Conversely, if a vector field satisfies these conditions, then, from (2.5),

$$\int_M |\pounds_X g|^2\, dV = 0,$$

from which

$$\pounds_X g = 0 \quad \text{in } M,$$

and consequently X is a Killing vector field; so we have

THEOREM 2.1. *A necessary and sufficient condition for a vector field X in M with boundary P to be a Killing vector field is that*

$$\begin{cases} \square X = 0, \quad \delta\xi = 0 & \textit{in } M \\ (\pounds_X g)(X, N) = 0 & \textit{on } P. \end{cases} \tag{2.6}$$

(Hsiung [3–6], Yano and Ako [2]).

If the vector field X vanishes on the boundary P, then the second condition of (2.6) is automatically satisfied; so we have

PROPOSITION 2.2. *A necessary and sufficient condition for a vector field X in M with boundary P vanishing on P to be a Killing vector field is that*

$$\Box X = 0, \qquad \delta\xi = 0 \quad \text{in } M.$$

If on the boundary P

$$X = BX' + \alpha N, \tag{2.7}$$

then we obtain formula (1.10), from which

$$(\pounds_X g)(X, N) = (\nabla\xi)(N, X) + h(X', X') + \alpha^2 p_1 \\ +2\alpha(\delta'\xi') - \delta'(\alpha\xi'). \tag{2.8}$$

So from Theorem 2.1 we get

PROPOSITION 2.3. *A necessary and sufficient condition for a vector field X in M with boundary P to be a Killing vector field is that*

$$\begin{cases} \Box X = 0, \qquad \delta\xi = 0 \quad \text{in } M, \\ (\nabla\xi)(N, X) + h(X', X') + \alpha^2 p_1 + 2\alpha(\delta'\xi') - \delta'(\alpha\xi') = 0 \quad \text{on } P. \end{cases} \tag{2.9}$$

If the vector field X is tangential to the boundary P, then $\alpha = 0$, and consequently from Proposition 2.3 we obtain

PROPOSITION 2.4. *A necessary and sufficient condition for a vector field X in M tangential to the boundary P to be a Killing vector field is that*

$$\begin{cases} \Box X = 0, \qquad \delta\xi = 0 \quad \text{in } M, \\ (\nabla\xi)(N, X) + h(X', X') = 0 \quad \text{on } P \end{cases} \tag{2.10}$$

(Yano [13]).

If the vector field X is normal to the boundary P, then $X' = 0$ and $X = \alpha N$. Consequently,

$$(\pounds_X g)(X, N) = \alpha(\pounds_X g)(N, N) \\ = 2\alpha[\alpha p_1 - \delta\xi]$$

by virtue of (1.8). Thus from Theorem 2.1 we get

PROPOSITION 2.5. *A necessary and sufficient condition for a vector field X in M normal to the boundary P to be a Killing vector field is that*

$$\begin{cases} \Box X = 0, \qquad \delta\xi = 0 \quad \text{in } M, \\ \alpha p_1 = 0 \quad \text{on } P \end{cases} \tag{2.11}$$

(Yano [13]).

Now, forming Eq. (1.4) − (2.5), we obtain

$$\int_M [R(X, X) - |\nabla X|^2 + \tfrac{1}{2}|\pounds_X g|^2 - (\delta\xi)^2]\, dV$$
$$= \int_P [h(X', X') + \alpha^2 p_1 + 2\alpha(\delta'\xi')]\, dS. \qquad (2.12)$$

On the other hand, for a Killing vector field X,

$$(\pounds_X g)(BY', BZ') = (\pounds_{X'} g')(Y', Z') - 2\alpha h(X', Y') = 0,$$

from which, by contraction,

$$\delta'\xi' = -\alpha p_1.$$

Thus, formula (2.12) becomes

$$\int_M [R(X, X) - |\nabla X|^2]\, dV = \int_P [h(X', X') + \alpha(\delta'\xi')]\, dS, \qquad (2.13)$$

from which we obtain

PROPOSITION 2.6. *If $R(X, X) \leq 0$, and if a Killing vector field X satisfies one of the following alternate sets of conditions of P,*

(*i*) $h(X', X') \geq 0, \quad \alpha = \text{const},$

(*ii*) $h(X', X') \geq 0, \quad \delta'\xi' = 0,$

(*iii*) $X' = 0,$

then

$$R(X, X) = 0, \qquad \nabla X = 0 \quad \textit{in } M,$$

and in cases (*i*) *and* (*ii*)

$$h(X', X') = 0 \quad \textit{on } P.$$

Thus, if $R(X, X)$ is negative-definite, there is no such vector field other than zero (Bochner [13]).

PROPOSITION 2.7. *If $R(X, X)$ is negative-definite and $h(X', X') \geq 0$, then there is no Killing vector tangential to the boundary other than zero.*

PROPOSITION 2.8. *If $R(X, X)$ is negative-definite, there is no Killing vector normal to the boundary other than zero.*

§3. Conformal Killing Vector Fields in Riemannian Manifolds with Boundary

A vector field X is said to be a conformal Killing vector field when it satisfies

$$\pounds_X g = 2\rho g \qquad (3.1)$$

for a particular function ρ. The function ρ above is found to be $-(1/n)(\delta\xi)$. Consequently, (3.1) can also be expressed as

$$\pounds_X g + \frac{2}{n}(\delta\xi)g = 0. \tag{3.2}$$

From (3.1),

$$(\pounds_X \nabla)(Y, Z) = (d\rho)(Y)Z + (d\rho)(Z)Y - (D\rho)\cdot g(Y, Z)$$

for arbitrary vector fields Y and Z, from which, by contraction,

$$\Box X + \frac{n-2}{n} D(\delta\xi) = 0. \tag{3.3}$$

By a straightforward computation we can prove

$$\langle \Box X + \frac{n-2}{n} D(\delta\xi), X\rangle - \frac{1}{2}\left| \pounds_X g + \frac{2}{n}(\delta\xi)\cdot g \right|^2 = \delta\left[i(X)(\pounds_X g) + \frac{2}{n}(\delta\xi)\xi \right], \tag{3.4}$$

which is valid for an arbitrary vector field X. Integrating both members of Eq. (3.4) over M and applying Stokes' theorem on the right-hand side, we get

$$\int_M \left[\langle \Box X + \frac{n-2}{n} D(\delta\xi)X, X\rangle - \frac{1}{2}\left| \pounds_X g + \frac{2}{n}(\delta\xi)\cdot g \right|^2 \right] dV = -\int_P \left[(\pounds_X g)(X, N) + \frac{2}{n}(\delta\xi)\xi(N) \right] dS. \tag{3.5}$$

Suppose that X is a conformal Killing vector field. Then,

$$\Box X + \frac{n-2}{n} D(\delta\xi) = 0 \quad \text{in } M,$$

and

$$(\pounds_X g)(X, N) + \frac{2}{n}(\delta\xi)\xi(N) = 0 \quad \text{on } P.$$

Conversely, if a vector field X satisfies these conditions, then, from (3.5),

$$\int_M \left| \pounds_X g + \frac{2}{n}(\delta\xi)\cdot g \right|^2 dV = 0,$$

from which

$$\pounds_X g + \frac{2}{n}(\delta\xi)\cdot g = 0 \quad \text{in } M,$$

and consequently X is a conformal Killing vector field. Thus we obtain

THEOREM 3.1. *A necessary and sufficient condition for a vector field X in M with boundary P to be a conformal Killing vector field is that*

$$\begin{cases} \square X + \dfrac{n-2}{n} D(\delta\xi) = 0 & \text{in } M, \\ (\pounds_X g)(X, N) + \dfrac{2}{n}(\delta\xi)\xi(N) = 0 & \text{on } P \end{cases} \tag{3.6}$$

(Ba [3–5], Hsiung [3–6], Yano and Ako [2]).

If the vector field X vanishes on the boundary, then the second condition of (3.6) is automatically satisfied. So we get

PROPOSITION 3.2. *A necessary and sufficient condition for an infinitesimal transformation X in M with boundary P leaving P pointwise invariant to be a conformal motion is that*

$$\square X + \frac{n-2}{n} D(\delta\xi) = 0 \quad \text{in } M.$$

If we put

$$X = BX' + \alpha N$$

on P, then, from (1.10),

$$\begin{aligned} &(\pounds_X g)(X, N) + \frac{2}{n}(\delta\xi)\xi(N) \\ &\quad = (\nabla\xi)(N, X) + h(X', X') + \alpha^2 p_1 - \frac{n-2}{n}\alpha(\delta\xi) + 2\alpha(\delta'\xi') - \delta'(\alpha\xi'). \end{aligned}$$

Thus from Theorem 3.1 we obtain

PROPOSITION 3.3. *A necessary and sufficient condition for a vector field X in M with boundary P to be a conformal Killing vector is that*

$$\begin{cases} \square X + \dfrac{n-2}{n} D(\delta\xi) = 0 & \text{in } M, \\ (\nabla\xi)(N, X) + h(X', X') + \alpha^2 p_1 & \\ \qquad - \dfrac{n-2}{n}\alpha(\delta\xi) + 2\alpha(\delta'\xi') - \delta'(\alpha\xi') = 0 & \text{on } P \end{cases} \tag{3.7}$$

(Yano and Ako [2]).

If the vector field X is tangential to the boundary P, then $\alpha = 0$, and so

PROPOSITION 3.4. *A necessary and sufficient condition for a vector field X in M tangential to the boundary P to be a conformal Killing vector field is that*

$$\begin{cases} \square X + \dfrac{n-2}{n} D(\delta\xi) = 0 & \text{in } M, \\ (\nabla\xi)(N, X) + h(X', X') = 0 & \text{on } P \end{cases} \tag{3.8}$$

(Yano [13]).

If the vector field X is normal to the boundary, then $X' = 0$ and $X = \alpha N$, and consequently

$$(£_X g)(N, X) + \frac{2}{n}(\delta\xi)\xi(N) = \alpha(£_X g)(N, N) + \frac{2}{n}\alpha(\delta\xi)$$

$$= 2\left[\alpha^2 p_1 - \frac{n-1}{n}\alpha(\delta\xi)\right]$$

by virtue of (1.8). Thus from Theorem 3.1 we get

PROPOSITION 3.5. *A necessary and sufficient condition for a vector field X in M normal to the boundary P to be a conformal Killing vector field is that*

$$\begin{cases} \square X + \dfrac{n-2}{n} D(\delta\xi) = 0 & \text{in } M, \\ \alpha^2 p_1 - \dfrac{n-1}{n}\alpha(\delta\xi) = 0 & \text{on } P \end{cases} \tag{3.9}$$

(Yano [13]).

Forming (1.4) − (3.5), we obtain

$$\int_M \left[R(X, X) - \frac{n-2}{n}\nabla_X(\delta\xi) - |\nabla\xi|^2 + \frac{1}{2}\left|£_X g + \frac{2}{n}(\delta\xi)g\right|^2\right] dV$$

$$= \int_P \left[h(X', X') + \alpha^2 p_1 - \frac{n-2}{n}\alpha(\delta\xi) + 2\alpha(\delta'\xi')\right] dS$$

or

$$\int_M \left[R(X, X) - \frac{2(n-1)}{n}|\nabla\xi|^2 + \frac{1}{2}\left|£_X g + \frac{2}{n}(\delta\xi)\cdot g\right|^2\right] dV$$

$$= \int_P [h(X', X') + \alpha^2 p_1 - 2\alpha(\delta'\xi')]\, dS.$$

On the other hand, for a conformal Killing vector field X,

$$\left[£_X g + \frac{2}{n}(\delta\xi)g\right](BY', BZ') = (£_{X'} g')(Y', Z') - 2\alpha p_1' + \frac{2}{n}(\delta\xi)g' = 0,$$

from which

$$\delta'\xi' = \alpha p_1 - \frac{n-1}{n}(\delta\xi).$$

Thus, for a conformal Killing vector field X, we obtain, from the above integral formula,

$$\int_M \left[R(X, X) - \frac{2(n-1)}{n} |\nabla\xi|^2 \right] dV$$
$$= \int_P \left[h(X', X') + \alpha(\delta'\xi') + \frac{n-1}{n}\alpha(\delta'\xi') \right] dS, \quad (3.10)$$

from which we get

PROPOSITION 3.6. *If $R(X, X) \leq 0$, and if a conformal Killing vector field X satisfies one of the following alternate sets of conditions on P,*

(*i*) $h(X', X') \geq 0, \quad \alpha = 0,$

(*ii*) $h(X', X') \geq 0, \quad \delta'\xi' + \dfrac{n-1}{n}\delta\xi = 0,$

(*iii*) $X' = 0, \quad \delta\xi = 0,$

then

$$R(X, X) = 0, \qquad \delta\xi = 0 \quad \textit{in } M,$$

and in cases (*i*) *and* (*ii*),

$$h(X', X') = 0 \quad \textit{on } P.$$

Thus, if $R(X, X)$ is negative-definite, there is no such vector field other than zero (Bochner [13], Yano and Ako [2]).

PROPOSITION 3.7. *If $R(X, X)$ is negative-definite and $h(X', X') \geq 0$, then there is no conformal Killing vector field tangential to the boundary other than zero.*

PROPOSITION 3.8. *If $R(X, X)$ is negative-definite, then there is no conformal Killing vector normal to the boundary such that $\delta\xi = 0$ on P other than zero.*

PROBLEMS

1. Let X be an arbitrary vector field in a Riemannian manifold and ξ its associated 1-form, that is,

$$\xi(Y) = g(X, Y)$$

for an arbitrary vector field Y. Prove that

$$\langle \Delta X, X\rangle - \tfrac{1}{2}|d\xi|^2 - (\delta\xi)^2 = -\delta[i(X)\,d\xi + (\delta\xi)\cdot\xi].$$

(See (1.3).)

2. Prove that

$$\langle \Box X, X\rangle - \tfrac{1}{2}|£_X g|^2 + (\delta\xi)^2 = \delta[i(X)(£_X g) + (\delta\xi)\cdot\xi].$$

(See (2.4).)

3. Prove that

$$\left\langle \Box X + \frac{n-2}{n} D(\delta\xi), X \right\rangle - \frac{1}{2}\left| £_X g + \frac{2}{n}(\delta\xi)\cdot g \right|^2 = \delta\left[i(X)(£_X g) + \frac{2}{n}(\delta\xi)\cdot\xi \right].$$

(See (3.4).)

CHAPTER

8

Harmonic Forms and Killing Tensor Fields in Riemannian Manifolds with Boundary

In this chapter we discuss harmonic forms, Killing tensor fields and conformal Killing tensor fields in manifolds with boundary, thereby extending the results of the previous chapters. The material is taken from Duff and Spencer [1], Hsiung [3–6], Takahashi [1], Yano [12], and Yano and Takahashi [1].

§1. Tangential and Normal Components of a p-form on the Boundary

Consider a p-form ω in a Riemannian manifold M with boundary P. We call the *tangential part* of ω on the boundary P the p-form

$$\omega(BX_1', BX_2', \ldots, BX_p')$$

and denote it by $t\omega$. Thus

$$(t\omega)(X_1', X_2', \ldots, X_p') = \omega(BX_1', BX_2', \ldots, BX_p'). \tag{1.1}$$

We call the *normal part* of ω on the boundary P the $(p-1)$-form

$$\omega(N, BX_2', \ldots, BX_p')$$

and denote it by $n\omega$. Thus

$$(n\omega)(X_2', \ldots, X_p') = \omega(N, BX_2', \ldots, BX_p'), \tag{1.2}$$

which is also written as

$$n\omega = t(N \lrcorner\, \omega). \tag{1.3}$$

If $n\omega = 0$, then the p-form ω is said to be *tangential* to the boundary P, and if $t\omega = 0$, ω is said to be *normal* to P.

Let θ be a $(p + 1)$-form and ω a p-form. Then,

$$t(\theta \llcorner\, \omega) = t(\theta) \llcorner\, t(\omega) - n(\theta) \llcorner\, n(\omega), \tag{1.4}$$

and

$$n(\theta \llcorner\, \omega) = \langle n(\theta), t(\omega) \rangle. \tag{1.5}$$

If θ and ω are both p-forms,

$$\langle \theta, \omega \rangle = \langle t\theta, t\omega \rangle + \langle n\theta, n\omega \rangle, \tag{1.6}$$

and in particular

$$|\omega|^2 = |t\omega|^2 + |n\omega|^2. \tag{1.7}$$

From (1.1) we find by covariant differentiation along the boundary that

$$\begin{aligned}(\nabla'_{X'}t\omega)(X'_1, X'_2, \ldots, X'_p) = (\nabla_{BX'}\omega)(BX'_1, BX'_2, \ldots, BX'_p)\\ + \omega(h(X', X'_1)N, BX'_2, \ldots, BX'_p)\\ + \omega(BX'_1, h(X', X'_2)N, \ldots, BX'_p) + \cdots\\ + \omega(BX'_1, BX'_2, \ldots, h(X', X'_p)N),\end{aligned}$$

by virtue of the equation of Gauss, that is,

$$\begin{aligned}(\nabla'_{X'}t\omega)(X'_1, X'_2, \ldots, X'_p) = (\nabla_{BX'}\omega)(BX'_1, BX'_2, \ldots, BX'_p)\\ + h(X', X'_1)(n\omega)(X'_2, \ldots, X'_p)\\ - h(X', X'_2)(n\omega)(X'_1, X'_3, \ldots, X'_p) - \cdots\\ - h(X', X'_p)(n\omega)(X'_2, X'_3, \ldots, X'_1).\end{aligned} \tag{1.8}$$

From (1.2) we find, by covariant differentiation along the boundary P,

$$\begin{aligned}(\nabla'_{X'}n\omega)(X'_2, \ldots, X'_p) = (\nabla_{BX'}\omega)(N, BX'_2, \ldots, BX'_p)\\ - \omega(BHX', BX'_2, \ldots, BX'_p)\\ + \omega(N, h(X', X'_2)N, \ldots, BX'_p) + \cdots\\ + \omega(N, BX'_2, \ldots, h(X', X'_p)N,\end{aligned}$$

by virtue of the equations of Gauss and Weingarten, that is,

$$\begin{aligned}(\nabla'_{X'}n\omega)(X'_2, \ldots, X'_p)\\ = (\nabla_{BX'}\omega)(N, BX'_2, \ldots, BX'_p) - (t\omega)(HX', X'_2, \ldots, X'_p).\end{aligned} \tag{1.9}$$

From (1.8), we find

$$d(t\omega) = t(d\omega). \tag{1.10}$$

From (1.8) we also find, by contraction with respect to X' and X'_1,

$$\begin{aligned}-(\delta t\omega)(X'_2,\ldots,X'_p) = {} & -(\delta\omega)(BX'_2,\ldots,BX'_p) - (\nabla_N\omega)(N, BX'_2,\ldots,BX'_p)\\ & + p_1(n\omega)(X'_2,\ldots,X'_p)\\ & - (n\omega)(HX'_2, X'_3,\ldots,X'_p) - \cdots\\ & - (n\omega)(X'_2, X'_3,\ldots,HX'_p).\end{aligned} \tag{1.11}$$

From (1.9),

$$\begin{aligned}(dn\omega)(X'_1, X'_2, \ldots, X'_p) = {} & -(nd\omega)(X'_1, X'_2, \ldots, X'_p)\\ & + t(\nabla_N\omega)\,(X'_1, X'_2, \ldots, X'_p)\\ & - (t\omega)(HX'_1, X'_2, \ldots, X'_p)\\ & + (t\omega)(HX'_2, X'_1, \ldots, X'_p) + \cdots\\ & + (t\omega)(HX'_p, X'_2, \ldots, X'_1).\end{aligned} \tag{1.12}$$

From (1.9) we also find, by contraction,

$$\delta n\omega = -n(\delta\omega). \tag{1.13}$$

Thus, for a p-form ω', if we put

$$\begin{aligned}(C\omega')(X'_1, X'_2, \ldots, X'_p) = {} & \omega'(HX'_1, X'_2, \ldots, X'_p)\\ & - \omega'(HX'_2, X'_1, \ldots, X'_p) - \cdots\\ & - \omega'(HX'_p, X'_2, \ldots, X'_1)\end{aligned} \tag{1.14}$$

and

$$\begin{aligned}&(C'\omega')(X'_1, X'_2, \ldots, X'_p)\\ &\qquad = p_1\omega'(X'_1, X'_2, \ldots, X'_p) - (C\omega')(X'_1, X'_2, \ldots, X'_p),\end{aligned} \tag{1.15}$$

we obtain, from (1.10), (1.11), (1.12), (1.13),

$$\begin{cases} d(t\omega) = t(d\omega), \\ \delta(t\omega) = t(\delta\omega) + n\nabla_N\omega - C'(n\omega), \end{cases} \tag{1.16}$$

$$\begin{cases} d(n\omega) = -n(d\omega) + t\nabla_N\omega - C(t\omega), \\ \delta(n\omega) = -n(\delta\omega). \end{cases} \tag{1.17}$$

We will employ the notation

$$\begin{cases} C_p(\omega') = \langle C\omega', \omega'\rangle, \\ C'_p(\omega') = \langle C'\omega', \omega'\rangle \end{cases} \tag{1.18}$$

in the sequel.

§2. Harmonic p-forms in Riemannian Manifolds with Boundary

A p-form ω is called a harmonic form when it satisfies

$$d\omega = 0, \qquad \delta\omega = 0. \tag{2.1}$$

Hence for a harmonic form,

$$\Delta\omega = 0. \tag{2.2}$$

On the other hand, for an arbitrary p-form ω,

$$\langle \delta\, d\omega, \omega \rangle - |d\omega|^2 = \delta(d\omega \llcorner \omega), \tag{2.3}$$

and

$$\langle d\, \delta\omega, \omega \rangle - |\delta\omega|^2 = -\delta(\omega \llcorner \delta\omega) \tag{2.4}$$

from which

$$\langle \Delta\omega, \omega \rangle - |d\omega|^2 - |\delta\omega|^2 = \delta(d\omega \llcorner \omega - \omega \llcorner \delta\omega). \tag{2.5}$$

Integrating both members of Eq. (2.5) and applying Stokes' theorem, we obtain

$$\int_M [\langle \Delta\omega, \omega \rangle - |d\omega|^2 - |\delta\omega|^2]\, dV = -\int_P \langle N, d\omega \llcorner \omega - \omega \llcorner \delta\omega \rangle\, dS\ . \tag{2.6}$$

But we know that

$$\begin{aligned} t(\langle N, d\omega \llcorner \omega \rangle) &= t(N \lrcorner (d\omega \llcorner \omega)) \\ &= n(d\omega \llcorner \omega), \end{aligned}$$

that is,

$$t(\langle N, d\omega \llcorner \omega \rangle) = \langle n\, d\omega, t\omega \rangle \tag{2.7}$$

and

$$\begin{aligned} t(\langle N, \omega \llcorner \delta\omega \rangle) &= t(N \lrcorner (\omega \llcorner \delta\omega)) \\ &= n(\omega \llcorner \delta\omega), \end{aligned}$$

that is,

$$t(\langle N, \omega \llcorner \delta\omega \rangle) = \langle n\omega, t\, \delta\omega \rangle, \tag{2.8}$$

by virtue of (1.3) and (1.5).

Thus, from (2.6),

$$\int_M [\langle \Delta\omega, \omega \rangle - |d\omega|^2 - |\delta\omega|^2]\, dV = -\int_P [\langle n\, d\omega, t\omega \rangle - \langle n\omega, t\, \delta\omega \rangle]\, dS\ . \tag{2.9}$$

Now, if ω is a harmonic p-form, then

$$\Delta\omega = 0 \quad \text{in } M$$

and

$$n\, d\omega = 0, \qquad t\, \delta\omega = 0 \quad \text{on } P.$$

Conversely, if these conditions are satisfied,

$$\int [|d\omega|^2 + |\delta\omega|^2]\, dV = 0,$$

from (2.9), so that

$$d\omega = 0, \qquad \delta\omega = 0,$$

which show that ω is harmonic. Hence

THEOREM 2.1. *A necessary and sufficient condition for a p-form ω in M with boundary P to be harmonic is*

$$\begin{cases} \Delta\omega = 0 & \text{in } M, \\ n\, d\omega = 0, \qquad t\, \delta\omega = 0 & \text{on } P \end{cases} \tag{2.10}$$

(Duff and Spencer [1], Hsiung [3–6], Takahashi [1]).

If ω is a harmonic p-form tangential to the boundary P, then

$$\Delta\omega = 0 \quad \text{in } M,$$

and

$$n\, d\omega = 0, \qquad n\omega = 0 \quad \text{on } P.$$

Conversely, if these conditions are satisfied, then, from (2.9), $d\omega = 0$ and $\delta\omega = 0$, and consequently ω is harmonic. Thus

PROPOSITION 2.2. *A necessary and sufficient condition for a p-form ω in M to be harmonic and tangential to the boundary P is*

$$\begin{cases} \Delta\omega = 0 & \text{in } M, \\ n\, d\omega = 0, \qquad n\omega = 0 & \text{on } P \end{cases} \tag{2.11}$$

(Takahashi [1], Yano [12], Yano and Takahashi [1]).

From the first equation of (1.17), we see that the condition

$$n\, d\omega = 0, \qquad n\omega = 0 \quad \text{on } P$$

is equivalent to

$$C(t\omega) = 0 \qquad n\omega = 0 \quad \text{on } P$$

Thus, from Proposition 2.2,

PROPOSITION 2.3. *A necessary and sufficient condition for a p-form ω in M to be harmonic and tangential to the boundary P is*

$$\begin{cases} \Delta\omega = 0 & \textit{in } M, \\ C(t\omega) = 0, \quad n\omega = 0 & \textit{on } P \end{cases} \tag{2.12}$$

(Takahashi [1], Yano [12], Yano and Takahashi [1]).

If ω is a harmonic p-form normal to the boundary P,

$$\Delta\omega = 0 \quad \text{in } M$$

and

$$t\omega = 0, \qquad t\,\delta\omega = 0 \quad \text{on } P.$$

Conversely, if these conditions are satisfied, then, from (2.9), $d\omega = 0$ and $\delta\omega = 0$, and consequently ω is harmonic. Hence

PROPOSITION 2.4. *A necessary and sufficient condition for a p-form ω in M to be harmonic and normal to the boundary P is*

$$\begin{cases} \Delta\omega = 0 & \textit{in } M, \\ t\omega = 0, \quad t\,\delta\omega = 0 & \textit{on } P \end{cases} \tag{2.13}$$

(Takahashi [1], Yano [12], Yano and Takahashi [1]).

From the second equation of (1.16), we see that the condition

$$t\omega = 0, \qquad t\,\delta\omega = 0 \quad \text{on } P$$

is equivalent to

$$t\omega = 0, \qquad C'(n\omega) = n(\nabla_N\omega) \quad \text{on } P.$$

Thus from Proposition 2.4 we obtain

PROPOSITION 2.5. *A necessary and sufficient condition for a p-form ω in M to be harmonic and normal to the boundary P is*

$$\begin{cases} \Delta\omega = 0 & \textit{in } M, \\ t\omega = 0, \quad C'(n\omega) = n(\nabla_N\omega) & \textit{on } P. \end{cases} \tag{2.14}$$

§3. Nonexistence of Harmonic p-forms

For a p-form ω we know that formula (2.5) holds:

$$\langle \Delta\omega, \omega\rangle - |d\omega|^2 - |\delta\omega|^2 = \delta(d\omega \llcorner \omega - \omega \llcorner \delta\omega). \tag{3.1}$$

On the other hand,

$$\langle \delta\,\nabla\omega, \omega\rangle - (p+1)\,|\nabla\omega|^2 = \delta(\nabla\omega \llcorner \omega). \tag{3.2}$$

Forming Eq. (3.1) − Eq. (3.2) and recalling the formula

$$\langle \Delta\omega - \delta\, \nabla\omega, \omega \rangle = F_p(\omega, \omega),$$

we find

$$F_p(\omega, \omega) + (p + 1)\, |\nabla\omega|^2 - |d\omega|^2 - |\delta\omega|^2$$
$$= \delta(d\omega \llcorner \omega - \omega \llcorner \delta\omega - \nabla\omega \llcorner \omega). \quad (3.3)$$

Integrating both members of Eq. (3.3) and applying Stokes' theorem on the left-hand member, we obtain

$$\int_M [F_p(\omega, \omega) + (p + 1)\, |\nabla\omega|^2 - |d\omega|^2 - |\delta\omega|^2]\, dV$$
$$= -\int_P \langle N, d\omega \llcorner \omega - \omega \llcorner \delta\omega - \nabla\omega \llcorner \omega \rangle\, dS\,. \quad (3.4)$$

But we know from (2.7) and (2.8) that

$$t(\langle N, d\omega \llcorner \omega \rangle) = \langle n\, d\omega, t\omega \rangle,$$
$$t(\langle N, \omega \llcorner \delta\omega \rangle) = \langle n\omega, t\, \delta\omega \rangle$$

and

$$t(\langle N, \nabla\omega \llcorner \omega \rangle) = \langle t(\nabla_N\omega), t\omega \rangle + \langle n(\nabla_N\omega), n\omega \rangle.$$

Consequently,

$$t(\langle N, d\omega \llcorner \omega - \omega \llcorner \delta\omega - \nabla\omega \llcorner \omega \rangle) = \langle n\, d\omega - t(\nabla_N\omega), t\omega \rangle$$
$$- \langle t\, \delta\omega + n(\nabla_N\omega), n\omega \rangle$$

or

$$t(\langle N, d\omega \llcorner \omega - \omega \llcorner \delta\omega - \nabla\omega \llcorner \omega \rangle) = -\langle C(t\omega), t\omega \rangle - \langle dn\omega, t\omega \rangle$$
$$- \langle C'(n\omega), n\omega \rangle - \langle \delta t\omega, n\omega \rangle$$
$$= -C_p(t\omega, t\omega) - C_p'(n\omega, n\omega)$$
$$- \langle dn\omega, t\omega \rangle - \langle \delta t\omega, n\omega \rangle.$$

Thus Eq. (3.4) becomes

$$\int_M [F_p(\omega, \omega) + (p + 1)\, |\nabla\omega|^2 - |d\omega|^2 - |\delta\omega|^2]\, dV$$
$$= \int_P [C_p(t\omega) + C_p'(n\omega) + \langle dn\omega, t\omega \rangle + \langle \delta t\omega, n\omega \rangle]\, dS$$

or

$$\int_M [F_p(\omega, \omega) + (p + 1)\, |\nabla\omega|^2 - |d\omega|^2 - |\delta\omega|^2]\, dV$$
$$= \int_P [C_p(t\omega) + C_p'(n\omega) + 2\langle dn\omega, t\omega \rangle]\, dS\,, \quad (3.5)$$

by virtue of

$$\int_P \langle dn\omega, t\omega\rangle \, dS = \int_P \langle n\omega, \delta t\omega\rangle \, dS ,$$

the boundary P being compact.

For a harmonic p-form ω we have, from (3.5),

$$\int_M [F_p(\omega, \omega) + (p + 1) |\nabla\omega|^2] \, dV = \int_P [C_p(t\omega) + C_p'(n\omega) + 2\langle dn\omega, t\omega\rangle] \, dS ,$$

from which

THEOREM 3.1. *If $F_p(\omega, \omega) \geq 0$ in M and if a harmonic p-form ω satisfies one of the following alternate sets of conditions,*

(i) $C_p(t\omega) \leq 0$, $n\omega = 0$,
(ii) $C_p(t\omega) \leq 0$, $C_p'(n\omega) \leq 0$, $dn\omega = 0$,
(iii) $C_p'(n\omega) \leq 0$ $t\omega = 0$,
on P, then

$$F_p(\omega, \omega) = 0, \qquad \nabla\omega = 0 \quad \textit{in } M$$

and

$$C_p(t\omega) = 0, \qquad C_p'(n\omega) = 0 \quad \textit{on } P.$$

Thus, if $F_p(\omega, \omega)$ is positive-definite, there exists no such p-form other than zero.

PROPOSITION 3.2. *If*
(i) $F_p(\omega, \omega)$ is positive-definite and $C_p(t\omega) \leq 0$ or
(ii) $F_p(\omega, \omega) \geq 0$ and $C_p(t\omega)$ is negative-definite, then there is no harmonic p-form tangential to the boundary other than zero.

PROPOSITION 3.3. *If*
(i) $F_p(\omega, \omega)$ is positive-definite and $C_p'(n\omega) \leq 0$, or
(ii) $F_p(\omega, \omega) \geq 0$ and $C_p'(n\omega)$ is negative-definite, then there is no harmonic p-form normal to the boundary other than zero.

Duff and Spencer [1] showed that the absolute pth Betti number $\ell_p(M)$ of M is equal to the number of linearly independent (with real coefficients) harmonic p-forms tangential to the boundary P, and the relative pth Betti number $\ell_p(M, P)$ of M modulo P is the number of linearly independent harmonic p-forms normal to the boundary P. Thus

PROPOSITION 3.4. *Under the assumption of Proposition* 3.2,

$$\ell_p(M) = \ell_{n-p}(M, P) = 0 \quad (0 < p < n).$$

PROPOSITION 3.5. *Under the assumption of Proposition* 3.3,

$$\ell_p(M, P) = \ell_{n-p}(M) = 0 \qquad (0 < p < n).$$

§4. Killing Forms in Riemannian Manifolds with Boundary

A p-form ω is called a *Killing p-form* when it satisfies

$$d\omega = (p+1)\,\nabla\omega, \tag{4.1}$$

and consequently

$$\delta\omega = 0. \tag{4.2}$$

Hence, for a Killing p-form ω,

$$\begin{aligned} \Delta\omega &= \delta\, d\omega + d\,\delta\omega \\ &= (p+1)\,\delta\,\nabla\omega, \end{aligned}$$

that is

$$\Box\omega = 0, \tag{4.3}$$

where

$$\Box\omega = \Delta\omega - (p+1)\,\delta\,\nabla\omega. \tag{4.4}$$

On the other hand, we have, for an arbitrary p-form ω,

$$\langle\delta\,\nabla\omega, \omega\rangle - (p+1)\,|\nabla\omega|^2 = \delta(\nabla\omega \llcorner \omega). \tag{4.5}$$

Thus, forming Eq. (2.5) $-$ $(p + 1) \times$ Eq. (4.5), we find

$$\begin{aligned} &\langle\Box\omega,\rangle\,\omega + |(p+1)\nabla\omega|^2 - |d\omega|^2 - |\delta\omega|^2 \\ &\qquad = \delta((d\omega - (p+1)\nabla\omega) \llcorner \omega - \omega \llcorner \delta\omega). \end{aligned}$$

But, since

$$\langle(p+1)\,\nabla\omega, d\omega\rangle = |d\omega|^2,$$

we have

$$|d\omega - (p+1)\,\nabla\omega|^2 = |(p+1)\,\nabla\omega|^2 - |d\omega|^2,$$

and consequently

$$\begin{aligned} &\langle\Box\omega, \omega\rangle + |d\omega - (p+1)\nabla\omega|^2 - |\delta\omega|^2 \\ &\qquad = \delta((d\omega - (p+1)\nabla\omega) \llcorner \omega - \omega \llcorner \delta\omega). \end{aligned} \tag{4.6}$$

Integrating both members of Eq. (4.6) and applying Stokes' theorem to the left-hand member,

$$\begin{aligned} &\int_M [\langle\Box\omega, \omega\rangle + |d\omega - (p+1)\,\nabla\omega|^2 - |\delta\omega|^2]\, dV \\ &\qquad = -\int_P \langle N, (d\omega - (p+1)\,\nabla\omega) \llcorner \omega - \omega \llcorner \delta\omega\rangle\, dS\,, \end{aligned} \tag{4.7}$$

which shows that if $\square\omega = 0$ and $\delta\omega = 0$ in M and $d\omega - (p+1)\nabla\omega = 0$ on P, then $d\omega - (p+1)\nabla\omega = 0$ is satisfied in M. Thus we have

THEOREM 4.1. *A necessary and sufficient condition for a p-form ω to be a Killing form is*

$$\begin{cases} \square\omega = 0, \qquad \delta\omega = 0 \quad \text{in } M, \\ d\omega = (p+1)\nabla\omega \quad \text{on } P. \end{cases} \tag{4.8}$$

Now, suppose that ω is a Killing p-form. Then it satisfies Eq. (4.8). But since

$$N \lrcorner (d\omega - (p+1)\nabla\omega) = N \lrcorner\, d\omega - (p+1)\nabla_N\omega$$

is a p-form on the boundary P, its tangential part is

$$n\, d\omega - (p+1)t(\nabla_N\omega)$$

and its normal part is

$$-(p+1)n(\nabla_N\omega).$$

So a Killing p-form ω satisfies

$$n\, d\omega - (p+1)t(\nabla_N\omega) = 0, \qquad n(\nabla_N\omega) = 0$$

on P.

Since

$$\begin{aligned} t\langle N, (d\omega - (p+1)\nabla\omega) \llcorner\, \omega - \omega \llcorner\, \delta\omega\rangle \\ &= n(d\omega \llcorner\, \omega) - (p+1)t(\nabla_N\omega \llcorner\, \omega) - n(\omega \llcorner\, \delta'\omega) \\ &= \langle n\, d\omega, t\omega\rangle - (p+1)[\langle t\,\nabla_N\omega, t\omega\rangle - \langle n\,\nabla_N\omega, n\omega\rangle] - \langle n\omega, t\,\delta\omega\rangle \\ &= \langle n\, d\omega - (p+1)t\,\nabla_N\omega, t\omega\rangle - \langle t\,\delta\omega - (p+1)n(\nabla_N\omega), n\omega\rangle, \end{aligned}$$

we have, from Eq. (4.7),

$$\begin{aligned} &\int_M [\langle\square\omega, \omega\rangle + |d\omega - (p+1)\nabla\omega|^2 - |\delta\omega|^2]\, dV \\ &\quad = -\int_P [\langle n\, d\omega - (p+1)t(\nabla_N\omega), t\omega\rangle - \langle t\,\delta\omega - (p+1)n(\nabla_N\omega), n\omega\rangle]\, dS \end{aligned} \tag{4.9}$$

Thus we have

THEOREM 4.2. *A necessary and sufficient condition for a p-form ω in M with boundary P to be a Killing form is*

$$\begin{cases} \square\omega = 0, \qquad \delta\omega = 0 \quad \text{in } M, \\ n\, d\omega = (p+1)t(\nabla_N\omega),\ n(\nabla_N\omega) = 0 \quad \text{on } P. \end{cases} \tag{4.10}$$

PROPOSITION 4.3. *A necessary and sufficient condition for a p-form ω in M with boundary P to be a Killing form tangential to P is*

$$\begin{cases} \Box\omega = 0, & \delta\omega = 0 \quad \text{in } M, \\ n\, d\omega = (p+1)t(\nabla_N\omega), & n\omega = 0 \quad \text{on } P. \end{cases} \tag{4.11}$$

On the other hand, from the first equation of (1.17),

$$n\, d\omega - (p+1)t(\nabla_N\omega) = -d(n\omega) - pt(\nabla_N\omega) - C(t\omega),$$

which shows that

$$n\, d\omega = (p+1)t(\nabla_N\omega), \qquad n\omega = 0$$

are equivalent to

$$pt(\nabla_N\omega) = -C(t\omega), \qquad n\omega = 0.$$

Thus from Proposition 4.3, we have

PROPOSITION 4.4. *A necessary and sufficient condition for a p-form ω in M with boundary P to be a Killing form tangential to P is*

$$\begin{cases} \Box\omega = 0, & \delta\omega = 0 \text{ in } M, \\ pt(\nabla_N\omega) = -C(t\omega), & n\omega = 0 \quad \text{on } P. \end{cases} \tag{4.12}$$

Now from (4.9) we have

PROPOSITION 4.5. *A necessary and sufficient condition for a p-form ω in M with boundary P to be a Killing form normal to P is*

$$\begin{cases} \Box\omega = 0, & \delta\omega = 0 \quad \text{in } M, \\ n(\nabla_N\omega) = 0, & t\omega = 0 \quad \text{on } P. \end{cases} \tag{4.13}$$

But by Eq. (1.16), $\Box\omega = 0$ and $\delta\omega = 0$ on M, the conditions $n(\nabla_N\omega) = 0$ and $t\omega = 0$ are equivalent to $C'(n\omega) = 0$ and $t\omega = 0$. Hence

PROPOSITION 4.6. *A necessary and sufficient condition for a p-form ω in M with boundary P to be a Killing form normal to P is*

$$\begin{cases} \Box\omega = 0, & \delta\omega = 0 \quad \text{in } M, \\ C'(n\omega) = 0, & t\omega = 0 \quad \text{on } P. \end{cases}$$

§5. Nonexistence of Killing p-forms

Since

$$\langle (p+1)\, \nabla\omega, d\omega \rangle = |d\omega|^2$$

for an arbitrary p-form ω, we have

$$|d\omega - (p+1)\nabla\omega|^2 = |(p+1)\nabla\omega|^2 - |d\omega|^2.$$

Consequently,

$$(p+1)|\nabla\omega|^2 - |d\omega|^2 = |d\omega - (p+1)\nabla\omega|^2 - p(p+1)|\nabla\omega|^2.$$

Thus Eq. (3.5) can be written as

$$\int_M [F_p(\omega,\omega) + |d\omega - (p+1)\nabla\omega|^2 - p(p+1)|\nabla\omega|^2 - |\delta\omega|^2]\, dV$$
$$= \int_P [C_p(t\omega) + C_p'(n\omega) + 2\langle dn\omega, t\omega\rangle]\, dS\,. \quad (5.1)$$

In particular, if ω is a Killing form, that is, if ω satisfies

$$d\omega = (p+1)\nabla\omega, \qquad \delta\omega = 0,$$

then Eq. (5.1) becomes

$$\int_M [F_p(\omega,\omega) - p(p+1)|\nabla\omega|^2]\, dV$$
$$= \int_P [C_p(t\omega) + C_p'(n\omega) + 2\langle dn\omega, t\omega\rangle]\, dS\,, \quad (5.2)$$

from which

THEOREM 5.1. *If $F_p(\omega,\omega) \le 0$ in M, and if a Killing p-form ω satisfies one of the following alternate sets of conditions,*

(*i*) $C_p(t\omega) \ge 0, \quad n\omega = 0,$
(*ii*) $C_p(t\omega) \ge 0, \quad C_p'(n\omega) \ge 0, \quad dn\omega = 0,$
(*iii*) $C_p'(t\omega) \ge 0, \quad t\omega = 0,$
on P, then

$$F_p(\omega,\omega) = 0, \qquad \nabla\omega = 0 \quad \text{in } M$$

and

$$C_p(t\omega) = 0, \qquad C_p'(n\omega) = 0 \quad \text{on } P.$$

Thus, if $F_p(\omega,\omega)$ is negative-definite, there exists no such p-form other than zero.

PROPOSITION 5.2. *If*
(*i*) *$F_p(\omega,\omega)$ is negative-definite and $C_p(t\omega) \ge 0$ or*
(*ii*) *$F_p(\omega,\omega) \le 0$ and $C_p(t\omega)$ is positive-definite, then there is no Killing p-form tangential to the boundary other than zero.*

PROPOSITION 5.3. *If*

(i) $F_p(\omega, \omega)$ *is negative-definite and* $C'_p(n\omega) \geq 0$, *or*

(ii) $F_p(\omega, \omega) \leq 0$ *and* $C'_p(n\omega)$ *is positive-definite, then there is no Killing p-form normal to the boundary other than zero.*

PROBLEMS

1. Let θ be a $(p + 1)$-form and ω a p-form. Prove that

$$t(\theta \llcorner \omega) = t(\theta) \llcorner t(\omega) - n(\theta) \llcorner n(\omega)$$

and

$$n(\theta \llcorner \omega) = \langle n(\theta), t(\omega)\rangle.$$

(See (1.4) and (1.5).)

2. Let θ and ω be p-forms. Prove that

$$\langle \theta, \omega\rangle = \langle t\theta, t\omega\rangle + \langle n\theta, n\omega\rangle$$

and, in particular,

$$|\omega|^2 = |t\omega|^2 + |n\omega|^2.$$

(See (1.6) and (1.7).)

3. Prove that for a p-form ω

$$\begin{cases} d(t\omega) = t(d\omega), \\ \delta(t\omega) = t(\delta\omega) + n\,\nabla_N\omega - C'(n\omega) \end{cases}$$

and

$$\begin{cases} d(n\omega) = -n(d\omega) + t\,\nabla_N\omega - C(t\omega), \\ \delta(n\omega) = -n(\delta\omega). \end{cases}$$

(See (1.16) and (1.17).)

4. Prove that for a p-form ω

$$\langle \delta\, d\omega, \omega\rangle - |d\omega|^2 = \delta(d\omega \llcorner \omega),$$

$$\langle d\,\delta\omega, \omega\rangle - |\delta\omega|^2 = -\delta(\omega \llcorner \delta\omega),$$

and consequently

$$\langle \Delta\omega, \omega\rangle - |d\omega|^2 - |\delta\omega|^2 = \delta(d\omega \llcorner \omega - \omega \llcorner \delta\omega).$$

(See (2.3), (2.4), and (2.5).)

5. Prove that for a p-form ω

$$t(\langle N, d\omega \llcorner \omega\rangle) = \langle n\, d\omega, t\omega\rangle$$

and

$$t(\langle N, \omega \llcorner \delta\omega\rangle) = \langle n\omega, t\,\delta\omega\rangle.$$

(See (2.7) and (2.8).)

6. Prove that for a p-form ω

$$\langle \delta\, \nabla\omega, \omega\rangle - (p + 1)\,|\nabla\omega|^2 = \delta(\nabla\omega \llcorner \omega),$$

and consequently

$$\langle \square\omega, \omega\rangle + |d\omega - (p+1)\nabla\omega|^2 - |\delta\omega|^2 = \delta((d\omega - (p+1)\nabla\omega \llcorner \omega - \omega \llcorner \delta\omega).$$

(See (4.5) and (4.6).)

7. Generalize the theorem in Problem 2 of Chapter 4 to the case of Riemannian manifolds with boundary (Takahashi [1]).
8. Generalize the theorem in Problem 3 of Chapter 4 to the case of Riemannian manifolds with boundary (Takahashi [1]).
9. Generalize the theorem in Problem 4 of Chapter 4 (Takahashi [1]).
10. Generalize the theorem in Problem 5 of Chapter 4 (Takahashi [1]).

Bibliography

Aeppli, A.

[1] Einige Ähnlichkeits- und Symmetriesatze für differenzierbare Flächen in Raum, *Comm. Math. Helv.* **33** (1959), 174–195.

Alexandrov, A. D.

[1] Ein allgemeiner Eindeutigkeitssatz für geschlossenen Flächen, *Compt. Rend. Acad. Sci. URSS* (Doklady) **19** (1938), 227–229.

[2] Uniqueness theorems for surfaces in the large, V, *Vestnik Leningrad Univ.* **13** (1958), 5–8.

[3] A characteristic property of spheres, *Ann. Mat. Pura Appl.* **58** (1962), 303–315.

Allendoerfer, C. B.

[1] The Euler number of a Riemannian manifold, *Am. J. Math.* **62** (1940), 243–248.

Auslander, L., and R. MacKenzie

[1] "Introduction to Differentiable Manifolds," McGraw-Hill, New York, 1963.

Ba, B.

[1]* Sur les transformations des variétés presque kählériennes, *Compt. Rend.* **252** (1961), 3719–3721.

[2]* Fonctions et champs de vecteurs presque complexes sur une variété presque complexe, *Compt. Rend.* **255** (1962), 819–821.

[3] Transformations conformes et presque analytiques des variétés presque kählériennes à bord, *Compt. Rend.* **257** (1963), 3554–3556.

[4] Sur les transformations conformes des variétés à bord, *Compt. Rend.* **258** (1964), 54–56.

[5] Structures presque complexes, structures conformes et dérivations, *Cahiers de Topologie et Géométrie Différentielle* **8** (1968), 1–74.

Barbance, C.

[1] Transformations conformes d'une variété riemannienne compacte, *Compt. Rend.* **260** (1965), 1547–1549.

Bishop, R. L., and R. J. Crittenden

[1] "Geometry of Manifolds," Academic Press, New York, 1964.

BISHOP, R. L., and S. I. GOLDBERG
[1] A characterization of the Euclidean sphere, *Bull. Am. Math. Soc.*, **72** (1966), 122–124.
[2] "Tensor Analysis on Manifolds," Macmillan, New York, 1968.
BOCHNER, S.
[1] Remarks on the theorem of Green, *Duke Math. J.* **3** (1937), 334–338.
[2] Vector field and Ricci curvature, *Bull. Am. Math. Soc.* **52** (1946), 776–797.
[3]* Curvature in Hermitian manifolds, *Bull. Am. Math. Soc.* **53** (1947), 179–195.
[4]* On compact complex manifolds, *J. Indian Math. Soc.* **116** (1947), 1–21.
[5] Curvature and Betti numbers, *Ann. Math.* **49** (1948), 379–390.
[6] Curvature and Betti numbers, II, *Ann. Math.* **50** (1949), 77–93.
[7] Euler-Poincaré characteristic for locally homogeneous and complex spaces, *Ann. Math.* **51** (1950), 241–261.
[8] Vector fields on complex and real manifolds, *Ann. Math.* **52** (1950), 642–649.
[9]* Complex spaces with transitive commutative groups of transformations, *Proc. Nat. Acad. Sci. U.S.* **37** (1951), 356–357.
[10] A new viewpoint in differential geometry, *Can. J. Math.* **3** (1951), 460–470.
[11]* Tensor fields and Ricci curvature in a Hermitian metric, *Proc. Nat. Acad. Sci. U.S.* **37** (1951), 704–706.
[12] Laplace operator on manifolds, *Proc. Intern. Congr. Mathematicians*, II (1952), 189–201.
[13] Vector fields on Riemannian spaces with boundary, *Annali di Mat.* **53** (1961), 57–62.
BOCHNER, S., and K. YANO
[1] Tensor fields in non-symmetric connections, *Ann. Math.* **56** (1952), 504–519.
BONNESEN, T., and W. FENCHEL
[1] "Theorie der konvexen Körper," Springer, Berlin, 1934.
BOOTHBY, W. M., and H. C. WANG
[1]* On contact manifolds, *Ann. Math.* **68** (1958), 721–734.
CARTAN, E.
[1] La géométrie des espaces de Riemann, Mémorial des Sci. Math., 1925.
[2] Leçons sur la Géométrie des Espaces de Riemann, Gauthier-Villars, Paris, 1928.
CHERN, S. S.
[1] On the curvature integral in a Riemannian manifold, *Ann. Math.* **46** (1945), 674–684.
[2] "Topics of Differential Geometry," Institute for Advanced Study, Princeton, 1951.
[3] Differentiable manifolds, Chicago lecture note, 1952.
[4] Some formulas in the theory of surfaces, *Bol. Soc. Mat. Mex.* **10** (1953), 30–40.
[5] Some new characterizations of the Euclidean sphere, *Duke Math. J.* **129** (1954), 270–290.
[6] On the curvature of a piece of hypersurface in Euclidean space, *Abh. Math. Seminar. Univ. Hamburg* **29** (1965), 77–91.
[7] Minimal submanifolds in a Riemannian manifold, Tech. Rep. 19, University of Kansas, 1968.
CHERN, S. S., M. do CARMO, and S. KOBAYASHI
[1] Minimal submanifolds of a sphere with second fundamental form of constant length, to appear.

COUTY, R.

[1] Vecteurs et tenseurs invariants sur un espace homogène, *Compt. Rend.* **246** (1958), 2569–2571.

[2] Transformations infinitésimales projectives, *Compt. Rend.* **247** (1958), 804–806.

[3] Sur les transformations des variétés riemanniennes et kählériennes, *Ann. Inst. Fourier (Grenoble)* **9** (1959), 147–248.

[4]* Transformations projectives des variétés presque kählériennes, *Compt. Rend.* **254** (1962), 4132–4135.

CRAIG, H. V.

[1] "Vector and Tensor Analysis," McGraw-Hill, New York, 1943.

DUFF, G. F. D., and D. C. SPENCER

[1] Harmonic tensors on Riemannian manifolds with boundary, *Ann. Math.* **56** (1952), 128–156.

EHRESMANN, C.

[1]* Sur les variétés presque complexes, *Proc. Intern. Congr. Mathematicians*, II (1950), 412–419.

EISENHART, L. P.

[1] "Riemannian Geometry," Princeton Univ. Press, Princeton, N.J., 1926; second printing, 1949.

[2] "Non-Riemannian Geometry," Am. Math. Soc. Colloq. Publ., Providence, R.I., 1927.

FLANDERS, H.

[1] "Differential Forms," Academic Press, New York, 1963.

GARDNER, R.

[1] An integral formula for immersions in Euclidean space. To appear in *J. Diff. Geom.*

GERRETSEN, J. C. H.

[1] "Lectures on Tensor Calculus and Differential Geometry," Noordhoff, Groningen, 1962.

GOLDBERG, S. I.

[1]* Conformal transformations of Kähler manifolds, *Bull. Am. Math. Soc.* **66** (1960), 54–58.

[2]* Groups of automorphisms of almost Kähler manifolds, *Bull. Am. Math. Soc.* **66** (1960), 180–183.

[3]* Groups of transformations of Kähler and almost Kähler manifolds, *Comm. Math. Helv.* **35** (1961), 35–46.

[4] "Curvature and Homology," Academic Press, New York, 1962.

[5] Manifolds admitting a one-parameter group of conformal transformations, *Mich. Math. J.* **15** (1968), 339–344.

GOLDBERG, S. I., and S. KOBAYASHI

[1] The conformal transformation group of a compact Riemannian manifold, *Proc. Nat. Acad. Sci. U.S.* **48** (1962), 25–26.

[2] The conformal transformation group of a compact homogeneous Riemannian manifold, *Bull. Am. Math. Soc.* **68** (1962), 378–381.

[3] The conformal transformation group of a compact Riemannian manifold, *Am. J. Math.* **84** (1962), 170–174.

GOLDBERG, S. I., and K. YANO

[1] Manifolds admitting a non-homothetic conformal transformation, to appear in Duke Math. J.

GRAY, T. W.
[1]* Some global properties of contact structures, *Ann. Math.* **69** (1959), 421–450.
GUGGENHEIMER, H.
[1]* Formes et vecteurs analytiques, *Ann. Mat. Pura Appl.* (4) **36** (1954), 223–246.
[2] "Differential Geometry," McGraw-Hill, New York, 1963.
HARDY, G. H., J. E. LITTLEWOOD, and G. POLYA
[1] "Inequalities," Cambridge Univ. Press, London, 1934.
HATAKEYAMA, Y.
[1]* On the existence of the Riemannian metric associated with a 2-form of rank $2r$, *Tôhoku Math. J.* **14** (1962), 162–166.
HELGASON, S.
[1] "Differential Geometry and Symmetric Spaces," Academic Press, New York, 1963.
HICKS, N.
[1] "Notes of Differential Geometry," Van Nostrand, Princeton, 1963.
HILT, A. L., and C. C. HSIUNG
[1] Vector fields and infinitesimal transformations on almost Hermitian manifolds with boundary, *Can. J. Math.* **17** (1965), 213–238.
HODGE, W. V. D.
[1] "Theory and Applications of Harmonic Integrals," Cambridge Univ. Press, London, 1941; 2nd ed., 1951.
HOPF, E.
[1] Elementäre Bemerkungen über die Lösungen partieller Differentialgleichungen zweiter Ordnungen vom elliptischen Typus, *Sitzber. Preuss. Akad. Wiss.* **19** (1927), 141–152.
HOPF, H.
[1] Über komplex-analytische Mannigfaltigkeit, *Rend. Mat. Appl.* **10** (1951), 169–182.
[2] Über Flächen mit einer Relation zwischen den Hauptkrümmangen, *Math. Nachr.* **4** (1951), 232–249.
HOPF, H., and Y. KATSURADA
[1] Some congruence theorems for closed hypersurfaces in Riemann spaces (Part II: Method based on a maximum principle), *Comm. Math. Helv.* **43** (1968), 217–223.
HOPF, H., and W. RINOW
[1] Über den Begriff der vollständigen differentialgeometrische Fläche, *Comm. Math. Helv.* **3** (1931), 209–225.
HOPF, H., and K. VOSS
[1] Ein Satz aus der Flächentheorie im Grossen, *Arch. Math.* **3** (1952), 187–192.
HSIUNG, C. C.
[1] Some integral formulas for closed hypersurfaces, *Math. Scand.* **2** (1954), 286–294.
[2] Some integral formulas for closed hypersurfaces in Riemannian space, *Pacific J. Math.* **6** (1956), 291–299.
[3] Curvature and Betti numbers of compact Riemannian manifolds with boundary, *Rend. Seminar. Mat. Univ. Politec. Torino* **17** (1957–1958), 95–131.
[4] A note of correction, *Rend. Seminar. Mat. Univ. Politec. Torino* **21** (1961–1962), 127–129.
[5] Vector fields and infinitesimal transformations on Riemannian manifolds with boundary, *Bull. Soc. Math. France* **92** (1964), 411–434.

[6] Curvature and homology of Riemannian manifolds with boundary, *Math. Z.* **82** (1963), 67–81.

[7] On the group of conformal transformations of a compact Riemannian manifold, *Proc. Natl. Acad. Sci. U.S.* **54** (1965), 1509–1513.

[8] On the group of conformal transformations of a compact Riemannian manifold, II, *Duke Math. J.* **34** (1967), 337–341.

[9] On the group of conformal transformations of a compact Riemannian manifold, III, *J. Diff. Geom.* **2** (1968), 185–190.

HSIUNG, C. C., and J. K. SHAHIN

[1] Affine differential geometry of closed hypersurfaces, *Proc. London Math. Soc.* (3), **17** (1967), 715–735.

HSIUNG, C. C., and J. D. LIU

[1] The group of conformal transformations of a compact Riemannian manifold, *Math. Z.* **105** (1968), 307–312.

ISHIHARA, S.

[1] On infinitesimal concircular transformations, *Kōdai Math. Sem. Rep.* **12** (1960), 45–56.

[2] The integral formulas and their applications in some affinely connected manifolds, *Kōdai Math. Sem. Rep.* **13** (1961), 93–108.

ISHIHARA, S., and Y. TASHIRO

[1] On Riemannian manifolds admitting a concircular transformation, *Math. J. Okayama Univ.* **9** (1959), 19–47.

KÄHLER, E.

[1]* Über eine bemerkenswerte Hermitische Metrik, *Abh. Math. Seminar. Hamburg* **9** (1933), 173–186.

KATSURADA, Y.

[1] Generalized Minkowski formulas for closed hypersurfaces in a Riemann space, *Annali di Mat.* **57** (1962), 283–294.

[2] On a certain property of closed hypersurfaces in an Einstein space, *Comm. Math. Helv.* **38** (1964), 165–171.

[3] On a piece of hypersurface in a Riemannian manifold with mean curvature bounded away from zero, *Trans. Am. Math. Soc.* **129** (1967), 447–457.

[4] Some congruence theorems for closed hypersurfaces in Riemann spaces (Part I: Method based on Stokes' theorem), *Comm. Math. Helv.* **43** (1968), 176–194.

[5] Closed submanifolds with constant ν-th mean curvature related with a vector field in a Riemannian manifold. *J. Fac. Sci. Hokkaido Univ.*, Ser. 1, **21** (1969), 171–181.

KATSURADA, Y., and H. KÔJYÔ

[1] Some integral formulas for closed submanifolds in a Riemann space, *J. Fac. Sci. Hokkaido Univ.*, Ser. 1, **20** (1968), 90–100.

KATSURADA, Y., and T. NAGAI

[1] On some properties of a submanifold with constant mean curvature in a Riemann space, *J. Fac. Sci. Hokkaido Univ.*, Ser. 1, **20** (1968), 79–89.

KILLING, W.

[1] Über die Grundlagen der Geometrie, *J. reine angew. Math.* **109** (1892), 121–186.

KOBAYASHI, S., and K. NOMIZU

[1] "The Foundations of Differential Geometry," Vol. I, Interscience Tracts in Pure and Applied Mathematics, Wiley-Interscience, New York, 1963.

[2] "The Foundations of Differential Geometry," Vol. II, Interscience Tracts in Pure and Applied Mathematics, Wiley-Interscience, New York, 1969.

KOSTANT, B.

[1] Holonomy and the Lie algebra of infinitesimal motions of a Riemannian manifold, *Trans. Am. Math. Soc.* **80** (1955), 528–542.

KOTŌ, S.

[1]* On harmonic tensors in an almost Tachibana space, *Tôhoku Math. J.* **13** (1961), 423–426.

KOYANAGI, T.

[1] On a certain property of a closed hypersurface in a Riemannian space, *J. Fac. Sci. Hokkaido Univ.*, Ser. 1, **20** (1968), 115–121.

KURITA, M.

[1] A note on conformal mappings of certain Riemannian manifolds, *Nagoya Math. J.* **21** (1962), 111–114.

LANG, S.

[1] "Introduction to Differentiable Manifolds," Wiley-Interscience, New York, 1962.

LAUGWITZ, D.

[1] "Differential and Riemannian Geometry," Academic Press, New York, 1965.

LEDGER, A. J., and M. OBATA

[1] Compact Riemannian manifolds with essential group of conformorphisms, to appear.

LEVI-CIVITA, T.

[1] "The Absolute Differential Calculus," Blackie, Glasgow, 1927.

LICHNEROWICZ, A.

[1] Courbure et nombres de Betti d'une variété riemannienne compacte, *Compt. Rend.* **226** (1948), 1678–1680.

[2] Courbure, nombres de Betti et espaces symmétriques, *Proc. Intern. Congr. of Mathematicians* **2** (1952), 216–223.

[3]* Sur les groupes d'automorphismes de certaines variétés kählériennes, *Compt. Rend.* **239** (1954), 1344–1347.

[4] Transformations infinitésimales conformes de certaines variétés riemanniennes compacts, *Compt. Rend.* **241** (1955), 726–729.

[5] "Théorie Globale des Connexions et des Groupes d'Holonomie," Ed. Cremonese, Rome, 1955.

[6] Some problems on transformations of Riemannian and Kählerian manifolds, Mimeographed notes, Princeton Univ., Princeton, 1955.

[7] Sur les transformations affines des variétés riemanniennes, *Compt. Rend.* **242** (1956), 1568–1570.

[8]* Sur les transformations analytiques des variétés kählériennes compactes, *Compt. Rend.* **244** (1957), 3011–3019.

[9]* Transformations analytiques et isométries d'une variété kählérienne compacte, *Compt. Rend.* **246** (1958), 855–857.

[10] "Géométrie des Groupes de Transformations," Dunod, Paris, 1958.

[11] Sur les transformations conformes d'une variété riemannienne compacte, *Compt. Rend.* **259** (1964), 697–700.

LIEBMANN, H.

[1] Über die Verbiegung der geschlossenen Flächen positiver Krümmung, *Math. Ann.* **53** (1900), 91–112.

Matsushima, Y.

[1] Sur la structure du groupe d'homéomorphismes analytiques d'une certaine variété kählérienne, *Nagoya Math. J.* **11** (1957), 145–150.

McConnell, A. J.

[1] "Applications of the Absolute Differential Calculus," Blackie, Glasgow, 1943.

Mogi, I.

[1] On harmonic fields in Riemannian manifolds, *Kōdai Math. Sem. Rep.* **26** (1950), 61–66.

Myers, S. B.

[1] Riemannian manifolds with positive mean curvature, *Duke Math. J.* **8** (1941), 401–404.

Nagai, T.

[1] On certain conditions for a submanifold in a Riemannian space to be isometric to a sphere, *J. Fac. Sci. Hokkaido Univ.*, Ser. 1, **21** (1969), 135–159.

Nagano, T.

[1] On conformal transformations of Riemannian spaces, *J. Math. Soc. Japan* **10** (1958), 79–93.

[2] The conformal transformations on a space with parallel Ricci curvature, *J. Math. Soc. Japan* **11** (1959), 10–14.

Nakae, T.

[1] Curvature and relative Betti numbers, *J. Math. Soc. Japan* **9** (1957), 367–373.

Newlander, A., and L. Nirenburg

[1]* Complex analytic coordinates in almost complex manifolds, *Ann. Math.* **65** (1957), 391–404.

Nijenhuis, A., and W. B. Woolf

[1]* Three integration problems in almost complex manifolds, *Tech. Rept.* **13** (1962), 1–114.

Nomizu, K.

[1] "Lie Groups and Differential Geometry," Publ. Math. Soc. Japan, 1956.

Nomizu, K., and B. Smyth

[1] A formula of Simons' type and hypersurfaces with constant mean curvature, to appear in *J. Diff. Geom.*

Obata, M.

[1] Certain conditions for a Riemannian manifold to be isometric with a sphere, *J. Math. Soc. Japan* **14** (1962), 333–340.

[2] Conformal transformations of compact Riemannian manifolds, *Illinois J. Math.* **6** (1962), 292–295.

[3] Conformal transformations in Riemannian manifolds (in Japanese), *Sūgaku Math. Soc. Japan* **14** (1963), 152–164.

[4] Riemannian manifolds admitting a solution of a certain system of differential equations, Proc. United States-Japan Seminar in Differential Geometry, Kyoto, Japan, 1965, pp. 101–114.

[5] Quelques inégalités intégrales sur une variété riemannienne compacte, *Compt. Rend.* **264** (1967), 123–125.

[6] Conformal transformations of Riemannian manifolds, to appear.

[7] Conformally flat Riemannian manifolds admitting a one-parameter group of conformal transformations, to appear.

OGAWA, Y.

[1] On C-harmonic forms in a compact Sasakian space, *Tôhoku Math. J.* **19** (1967), 267–296.

OKUMURA, M.

[1] Compact orientable submanifolds of codimension 2 in an odd dimensional sphere, *Tôhoku Math. J.* **20** (1968), 8–20.

[2] Submanifolds of codimension 2 with certain properties, to appear.

O'NEILL, B.

[1] "Elementary Differential Geometry," Academic Press, New York, 1966.

ŌTSUKI, T.

[1] Integral formulas for hypersurfaces in a Riemannian manifold and their applications, *Tôhoku Math. J.* **17** (1965), 335–348.

DE RHAM, G.

[1] Sur la théorie des formes différentielles harmoniques, *Ann. Univ. Grenoble* **22** (1946), 135–156.

[2] Sur la réductibilité d'un espace de Riemann, *Comm. Math. Helv.* **26** (1952), 328–344.

[3] "Variétés Différentiables," Actual. Sci., Hermann, Paris, 1955.

DE RHAM, G., and K. KODAIRA

[1] "Harmonic Integrals," Institute for Advanced Study, Princeton, 1950.

SASAKI, S., and K. YANO

[1] Pseudo-analytic vectors on pseudo-Kählerian manifolds, *Pacific J. Math.* **5** (1955), 989–993.

SATŌ, I.

[1] On conformal Killing tensor fields, *Bull. Yamagata Univ.* **3** (1956), 175–180.

SAWAKI, S.

[1]* On Matsushima's theorem in a compact Einstein K-space, *Tôhoku Math. J.* **13** (1961), 455–465.

[2]* A generalization of Matsashima's theorem, *Math. Ann.* **146** (1962), 279–286.

SCHOUTEN, J. A.

[1] "Ricci-Calculus" (2nd ed.), Springer, Berlin, 1954.

SHAHIN, J. K.

[1] Some integral formulas for closed hypersurfaces in Euclidean space, *Proc. Am. Math. Soc.* **19** (1968), 609–613.

SIMONS, J.

[1] Minimal varieties in Riemannian manifolds, *Ann. Math.* **88** (1968), 62–105.

SOKOLNIKOFF, I.

[1] "Tensor Analysis," Wiley, New York, 1964.

SPAIN, B.

[1] "Tensor Calculus," Wiley-Interscience, New York, 1953.

SPRINGER, C. E.

[1] "Tensor and Vector Analysis with Application to Differential Geometry," Ronald, New York, 1962.

STERNBERG, S.

[1] "Lectures on Differential Geometry," Prentice-Hall, Englewood Cliffs, N.J., 1964.

STONG, D. J.

[1] Some differential geometric properties of submanifolds of Euclidean spaces, *Proc. Amer. Math. Soc.* **12** (1961), 343–349.

STRUIK, D. J.

[1] "Differential Geometry" (2nd ed.), Addison-Wesley, Reading, Mass., 1961.

SÜSS, W.

[1] Zur relativen Differentialgeometrie, V., *Tôhoku Math. J.* **30** (1929), 202–209.

SYNGE, J. L., and A. SCHILD

[1] "Tensor Calculus," Univ. of Toronto Press, Toronto, Canada, 1949.

TACHIBANA, S.

[1] On harmonic tensors in compact Sasakian spaces, *Tôhoku Math. J.* **17** (1965), 271–284.

TAKAHASHI, T.

[1] On harmonic and Killing vector fields in a Riemannian manifold with boundary, *J. Math. Soc. Japan* **14** (1962), 37–65.

TANI, M.

[1] On hypersurfaces with constant k-th mean curvature, *Kōdai Math. Sem. Rep.* **20** (1968), 94–102.

[2] On a piece of hypersurface in a fibred space, to appear.

TASHIRO, Y.

[1] Complete Riemannian manifolds and some vector fields, *Trans. Am. Math. Soc.* **117** (1965), 251–275.

TAZAWA, Y.

[1] A remark on a closed orientable hypersurface with constant reduced mean curvature, *J. Fac. Sci. Hokkaido Univ.*, Ser. 1, **20** (1968), 101–108.

THOMAS, T. Y.

[1] Some applications of Green's theorem for compact Riemannian spaces, *Tôhoku Math. J.* **46** (1940), 261–266.

[2] "Concepts from Tensor Analysis and Differential Geometry," Academic Press, New York, 1961.

VEBLEN, O.

[1] Invariants of quadratic differential forms, Cambridge Univ. Press, London, 1927.

VOSS, K.

[1] Einige differentialgeometrische Kongruenzsätze für geschlossene Flächen und Hyperflächen, *Math. Ann.* **131** (1956), 180–218.

WATANABE, Y.

[1] Integral inequalities in compact orientable manifolds, Riemannian or Kählerian, *Kōdai Math. Sem. Rep.* **20** (1968), 261–271.

WEATHERBURN, C. E.

[1] "An Introduction to Tensor Calculus and Riemannian Geometry," Cambridge Univ. Press, London, 1942.

WEBER, W. C., and S. I. GOLDBERG

[1] Conformal deformations of Riemannian manifolds, "Queen's Papers on Pure and Applied Mathematics," Queen's Univ., Kingston, Canada.

WEYL, H.

[1] Reine Infinitesimalgeometrie, *Math. Z.* **26** (1918), 384–411.

[2] "Raum, Zeit, Materie," Springer, Berlin, 1921.
[3] Zur Infinitesimalgeometrie: Einordnung der projektiven und der konformen Auffassung, *Göttingen Nachr.* 1921, 99–112.

WILLMORE, T. J.
[1] "An Introduction to Differential Geometry," Clarendon Press, Oxford, 1959.

YAMABE, H.
[1] On a deformation of Riemannian structures on compact manifolds, *Osaka Math. J.*, **12** (1960), 21-37.

YANO, K.
[1] Concircular geometry, I. Concircular transformations, *Proc. Imp. Acad. Japan* **16** (1940), 195–200.
[2] Concircular geometry, II. Integrability conditions of $\rho_{\mu\nu} = \phi g_{\mu\nu}$, *Proc. Imp. Acad. Japan* **16** (1940), 345–360.
[3] Concircular geometry, III. Theory of curves, *Proc. Imp. Acad. Japan* **16** (1940), 442–448.
[4] Concircular geometry IV. Theory of subspaces, *Proc. Imp. Acad. Japan* **16** (1940), 505–511.
[5] Concircular geometry, V. Einstein spaces, *Proc. Imp. Acad. Japan* **18** (1942), 446–451.
[6] On harmonic and Killing vector fields, *Ann. Math.* **55** (1952), 38–45.
[7] Some remarks on tensor fields and curvature, *Ann. Math.* **55** (1952), 328–341.
[8]* Sur un théorème de M. Matsushima, *Nagoya Math. J.* **12** (1957), 167–170.
[9] "The Theory of Lie Derivatives and Its Applications," North-Holland Publ. Co., Amsterdam, 1957.
[10]* Some integral formulas and their applications, *Mich. J. Math.* **5** (1958), 63–73.
[11] Sur les vecteurs harmoniques et vecteurs de Killing dans un espace de Riemann à frontière, *Compt. Rend.* **247** (1958), 1085–1087.
[12] Harmonic and Killing tensor fields in Riemannian spaces with boundary, *J. Math. Soc. Japan* **10** (1958), 430–437.
[13] Harmonic and Killing vector fields in compact orientable Riemannian spaces with boundary, *Ann. Math.* **69** (1959), 588–597.
[14] Conformal transformations in Riemannian and Hermitian spaces, *Bull. Am. Math. Soc.* **66** (1960), 369–372.
[15]* "Differential Geometry on Complex and Almost Complex Spaces," Pergamon Press, New York, 1965.
[16] Closed hypersurfaces with constant mean curvature in a Riemannian manifold, *J. Math. Soc. Japan* **17** (1965), 333–340.
[17] On Riemannian manifolds with constant scalar curvature admitting a conformal transformation group, *Proc. Nat. Acad. Sci. U.S.* **55** (1966), 472–476.
[18] Notes on hypersurfaces in a Riemannian manifold, *Can. J. Math.*, **19** (1967), 439–445.
[19] Riemannian manifolds admitting a conformal transformation group, *Proc. Nat. Acad. Sci. U.S.* **62** (1969), 314–319.
[20] On Riemannian manifolds admitting an infinitesimal conformal transformation, *Math. Z.* **113** (1970), 205–214.
[21] Integral formulas for submanifolds and their applications, to appear in *Can. J. Math.*

[22] Notes on submanifolds in a Riemannian manifold, *Kōdai Math. Sem. Rep.* **21** (1969), 496–509.

Yano, K., and M. Ako

[1]* Almost analytic vectors in almost complex spaces, *Tôhoku Math. J.* **13** (1961), 24–45.

[2] Vector fields in Riemannian and Hermitian manifolds with boundary, *Kōdai Math. Sem. Rep.* **17** (1965), 129–157.

Yano, K., and S. Bochner

[1] "Curvature and Betti numbers," *Ann. Math. Studies* **32** (1953).

Yano, K., and S. Ishihara

[1] Submanifolds with parallel mean curvature vector, to appear.

Yano, K., and T. Nagano

[1] Einstein spaces admitting a one-parameter group of conformal transformations, *Ann. Math.* **69** (1959), 451–461.

[2] On geodesic vector fields in a compact orientable Riemannian space, *Comm. Math. Helv.* **35** (1961), 55–64.

Yano, K., and M. Obata

[1] Sur le groupe de transformations conformes d'une varieté de Riemann dont le scalaire de courbure est constant, *Compt. Rend.* **260** (1965), 2698–2700.

[2] On conformal changes of Riemannian metrics, to appear in *J. Differential Geometry.*

Yano, K., and M. Okumura

[1] Integral formulas for submanifolds of codimension 2 and their applications, *Kōdai Math. Sem. Rep.* **21** (1969), 463–471.

Yano, K., and S. Sawaki

[1] On Riemannian manifolds admitting a conformal transformation group, *J. Differential Geometry* **2** (1968), 161–184.

[2] Riemannian manifolds admitting an infinitesimal conformal transformation, to appear.

Yano, K., and T. Takahashi

[1] Some remarks on harmonic and Killing tensor fields in a Riemannian manifold with boundary, Calcutta Math. Soc., The Golden Jubilee Commemoration Volume, 1958–1959, 439–446.

Yano, K., and M. Tani

[1] Integral formulas for closed hypersurfaces, *Kōdai Math. Sem. Rep.* **21** (1969), 335–349.

[2] Submanifolds of codimension 2 in a Euclidean space, to appear.

Author Index

Numbers in italics show the page on which the complete reference is listed.

V

W

Y

Subject Index